Without Trace

Without Trace

The last voyages of eight ships

John Harris

A Methuen Paperback

First published 1981 by Eyre Methuen Ltd
Reprinted 1982, 1983, 1987
This paperback edition published 1987
by Methuen London Ltd,
11 New Fetter Lane, London EC4P 4EE
© 1981 John Harris

Printed in Great Britain by
Richard Clay Ltd, Bungay, Suffolk

British Library Cataloguing in Publication Data

Harris, John, *1916–*
    Without trace: the last voyages
    of eight ships.
    1. Ocean
    I. Title
    001.9′4′09162   GC21

ISBN 0-413-15600-1

CONTENTS

ILLUSTRATIONS

Sir John Franklin (National Portrait Gallery, London)

Lady Franklin (Queen Victoria Museum, Launceston, Tasmania)

H.M.S. *Erebus* and *Terror* (Scott Polar Research Institute, Cambridge)

The message found by Captain McClintock (*Illustrated London News*)

"They forged the last link with their lives" (National Maritime Museum, London)

Sir Robert McClure (National Portrait Gallery, London)

The discovery of *Mary Celeste* (Atlantic Mutual Insurance Company, New York)

Mary Celeste in full sail (Peabody Museum, Salem, Massachusetts)

Captain Morehouse of *Dei Gratia* (Peabody Museum, Salem, Massachusetts)

Maine's wreck (Peabody Museum, Salem, Massachusetts)

Captain Charles Sigsbee (U.S. Navy)

William Randolph Hearst (BBC Hulton Picture Library)

The liner *Waratah* (P & O Archives)

Waratah's captain, chief officer and crew (La Trobe Library, Melbourne)

The U.S.S. *Cyclops* (Peabody Museum, Salem, Massachusetts)

Joyita as she was found (Ministry of Information, Fiji)

Joyita being towed to Suva (Ministry of Information, Fiji)

Donald Crowhurst leaving Teignmouth (*The Sunday Times*, London)

Teignmouth Electron being towed into Santo Domingo (Associated Press Ltd.)

MAPS

ACKNOWLEDGMENTS

Among others, my thanks are due to Commander John Manning, RN (retd.), who did a lot of the initial research; to the staffs of Durban and Cape Town newspapers, who responded at length to my requests for information, particularly Mr. George Young, Shipping Editor of the *Cape Times;* Mr. James R. Henderson III, of the *Virginian-Pilot and Norfolk Landmark,* Norfolk, Virginia, who sent me information on *Cyclops;* Rear Admiral Kemp Tolley, USN (retd.), who made available to me many items from the United States Naval Institute *Proceedings* and the detailed work by Admiral Rickover; and to the National Maritime Museum Library.

Without Trace

Introduction

A place where legends grow

When the five-masted schooner *Carroll A. Deering* was sighted on January 31, 1921, aground on the Diamond Shoals, off Cape Hatteras, North Carolina, with all sails set despite the stormy weather and with not a living thing on board save two cats, it started one of the classic mysteries of the sea.

No distress signals had been sighted. She had been hailed by the Cape Lookout lightship as she had passed three days before, still apparently in good order. But when examined as a derelict it appeared that her master, Captain Willis Wormell, had been doing the navigating only up to a certain point in her voyage, after which, judging by the writing on the chart, someone else had taken over. Though there were signs that a meal had been in preparation when the crew left, the forecastle was almost bare of baggage and clothing, the captain's trunk had gone and his cabin showed signs of having been occupied by some other person.

Messages in bottles were washed ashore and piracy was suggested, as also was hijacking, then particularly prevalent, with Prohibition in the United States and gangsters lifting each other's loads of illicit booze. Rumors also spread that Bolshevik agents had attacked the ship: with the Russian Revolution only four years past, Americans were assuming that democracy would vanish overnight unless all potential Bolsheviks were under lock and key and, though not one spy, "revolutionary workman" or Bolshevik was convicted of an overt act, intolerance, bigotry and hysteria were sweeping the country. Many were convinced that *Carroll A. Deering*'s crew had disappeared to a Russian port.

Nothing was ever settled and perhaps the only sensible thing that came out of the affair was the comment of the schooner's former master, Captain William Merritt—"It is so hard to prove anything that happens at sea."

This truth has been repeatedly endorsed by the surprising number of unsolved sea mysteries. When whole crews have been lost, it has always been impossible to prove exactly what happened. With just one survivor, there was no mystery.

In previous ages, the odds were often against sailors and even the more seaworthy ships of the nineteenth century were all too often at the mercy of the wind and the waves. Though technological advances have made the sea safer, ships still continue to be lost, sometimes vanishing without trace or apparent reason.

Sailors have always been superstitious. Until recently the objections to sailing on a Friday or to having a woman on board were strong and the belief was widespread that seagulls were the spirits of drowned men. As a result there are endless supernatural sea mysteries, some of which have a curiously convincing ring about them. To landsmen the sea is strange and sinister, and the men who follow it have an obvious mystique, from the saltiest admiral to the youngest pink-faced deck boy. To landsmen the sea is a mysterious element which is always likely to produce an enigma. With water covering the largest part of the earth's surface, there are boundless empty distances which have produced a whole crop of legends, mysteries and disasters.

It is a history of lost continents, piracy, hallucinations, treasure, murder, mutiny, madness and even cannibalism experienced by men in the extremities of suffering or loneliness. There are lost expeditions and vanishing ships, ghost crews and ghost vessels, and the spirits of the dead returning to the land of the living. There are monsters, political and financial chicanery, and many unexplained events, so it is little wonder that seamen felt so strongly about jinxed ships and preferred to avoid anything tainted by ill-luck.

The stories are legion. The author himself experienced the case of a seaman who knew he was going to die—and did, and in South Africa he even found himself involved in a ghost story. While operating during World War II from a whaling station which had been deserted since 1923, he was one of a group of men who arrived on the station in separate small parties, to every one of which there was a bad accident on the day of its arrival. One man fell through a jetty, another man cut his foot badly, a third fell from rocks and the author was washed overboard and badly injured his knee. It was only after this that he heard the story of the jinx that lay over the station. An elderly man who operated a small schooner in the vicinity had gone to sea as a boy from there. He told the author how at the end of the last century, at a time when the demand for whale oil was increasing and pressure was mounting for whalers to be at sea instead of alongside, he had been taken off his ship with a fever just before it sailed. While sleeping at the station one night he heard the ghostly voices of his crew who, it turned out, were already drowned. From then on, as if the station were warning them to take care, other ships lying alongside lost individual crew members by drowning, accident or murder. Eventually, with the arrival of factory ships, the station was closed and remained uninhabited until the author's group arrived there in late 1942. There was a remarkable similarity between the accidents which had happened in the station's heyday and those which befell the author's group. Was there a connection? Certainly four men in 1942 received injuries in a curious sequence of accidents.

It is true that ghost ships seem to have disappeared with the advent of the electric light. Perhaps the stories sprang from superstitious sailors who passed their night watches in a world of illusory shadows and flickering lanterns. Though he might be terrified by the phantoms he saw, the old-time sailor was never surprised, and, after all, there is the well-authenticated story of UB-65, the haunted German submarine in World War I, and stories of ghost ships in the Seine, the Solway Firth, the Gulf of St. Lawrence and Long Island Sound. Even that hardheaded around-the-world sailor Joshua Slocum claimed that during his circumnavigation, when he was lying desperately ill, he was saved by a mysterious and ghostly visitor who took over the running of his boat.

Because of the size of the sea, because of the sailors' habit of yarning in the dog watches and because of their tendency to improve on their stories when ashore for the chance of another drink, the sea has always been a place where legends grow. Some are extremely dubious. The Flying Dutchman, the story of the sea captain known variously as Cornelius Vanderdecken, van der Straaten or even the buccaneer, Soertebeker—who was condemned by God to sail perpetually around the Cape of Good Hope in a ghost ship—could be explained by St. Elmo's fire, the strange natural electric phenomenon that brushes off the tops of ships' masts, or by the mirages that occur off that coast. The author himself has seen ships there in positions where they couldn't possibly have been and experienced the strange steep seas of the East African coast. But the legends persist, like that of Atlantis, a vast continent said to have existed in the mid-Atlantic in the days before Christ and to have been destroyed by an earthquake, or that of the Sargasso Sea, a waste of floating seaweed off the coast of Florida which was believed to be the graveyard of hundreds of ships. After Columbus had described its haunting atmosphere, it became an obsession with seamen. The masses of weed produced tales to chill the blood of cabin boys, and merchant ships were said to have been strangled by monstrous growths of vegetation, their crews dying of hunger and thirst.

Every sailor knows the stories and many of them even have a logical explanation. Yet the Flying Dutchman was said to have been seen quite distinctly in 1881 from the brig *Bacchante*, among whose passengers was a young man who later became King George V.

Mermaids fascinated sailors. Usually they were alluring girls with long hair and fishes' tails. The ancient Greeks worshipped

sea nymphs, and the Romans believed that Sirens lived along the
Italian coasts, like seabirds on nests made of the bones of
drowned sailors. They never failed to catch the imagination.
Ulysses reported seeing them and Gerard Manley Hopkins wrote
of them. Edgar Allan Poe's *Annabelle Lee* is about a maiden
"from a kingdom 'neath the sea." Heine wrote about the Sirens
who lured vessels ashore from beneath the rock of the Lorelei.
Even Shakespeare did not forget them. Stories about them exist
in Iona and Orford in Suffolk. Columbus and Henry Hudson reported them and, in 1403, one was said to have been captured at
Edam, in Holland, which was taught to kneel before the crucifix
and lived in Haarlem for 15 years. They have been seen off Newfoundland and in the West Indies, and a respectable God-fearing
schoolmaster of Thurso claimed to have seen one in 1797. When
he recounted the story in 1809, neighbors also claimed they had
seen the mermaid. In fact these unnatural sightings could have
been of seals or walruses; of dugongs or manatees, cetacean
mammals of singular ugliness which, when suckling young,
might well have looked roughly like women. Perhaps the stories
were expanded by sex-starved sailors eager to improve a good
yarn.

No one who hasn't sailed the Pacific or the South Atlantic can
have any notion of the emptiness of the ocean or the size of the
denizens of those vast wastes—whales, giant squid, enormous
sharks, albatrosses and frigate birds, even living creatures thought
to have been extinct for thousands of years. There are stories of
monsters so big that they topped the masts of ships. Vast polyps
were reported on several occasions even in the last century, one
between Madeira and Tenerife, others off the coast of Newfoundland. In several cases severed tentacles, chopped off by alert
members of the crew when the monster attacked a ship, were
brought to port. One half tentacle was 18 feet long, which made
the octopus roughly 80 feet wide; and the schooner *Pearl* was
supposed to have been dragged under in 1874 by a giant squid—
though this story has to be taken with a grain of salt as the invention of a careless captain trying to explain away the loss of his ship.

The traditional sea serpent was often seen. In the last century one of them seemed to cross and recross the Atlantic between Scotland and New England. But then, in 1840, one monster, which had been seen on several occasions, was found to be nothing but a monstrous alga, over 100 feet long and four feet in diameter, whose root at a distance resembled a head and was given motion by the waves to make it seem alive. Nevertheless others swore that what they saw were *not* algae. Now the Loch Ness monster is being accepted by more and more naturalists, and it is a fact that the coelacanth has survived the passage of millions of years when it was previously believed to be extinct. Sometimes the people who saw the monsters were so close that it is unlikely they made a mistake. In 1875 a group on the yacht *Princess* followed their particular monster for two hours and managed to get close enough to look into its mouth. Another monster was seen more than once from the yacht *Valhalla*, owned by the Earl of Crawford, off the coast of Brazil at a point where the water was as deep as 1300 fathoms, by two Fellows of the Royal Zoological Society who surely ought to have known what they were talking about. It seems unlikely that all the witnesses were lying, especially as they were prepared to swear on the Bible in an era when Bible oaths were taken seriously. They were also certainly not all drunk and as sailors who knew what seals, sharks, porpoises and squid looked like, they were unlikely to be deceived.

It is possible that strange creatures could have survived in the deep trenches in the ocean floor. Certainly off South Africa the author saw a shark near his deserted whaling station so big that it seemed at first to be a whale. He was on an iron boat and was near enough to jab at it with a boathook. Later, off the west coast of Africa, he saw, again at a distance of only a few feet, what appeared to be two coils of a sea serpent. Imagining it to be a log washed down by the rains and likely to be of danger to the flying boats with which he was then operating, he went close to attach a rope and tow it away, when the two "coils" vanished, in a way that seemed to suggest they were attached. He saw the same thing again some days later and so did others. The "coils" might have been two whale sharks, enormous creatures that are virtually harmless—the color and size were right—but whale sharks have dorsal fins and the humps the author saw had no dorsal fins.

With the arrival of radio, the improved education of sailors and the increased speed of ships, which cuts the length of voyages, the scale and variety of the mysteries have diminished. But in any case most have become mysteries only with the passage of years and the attention of writers. For many so-called "mysteries" there existed a reasonable explanation, and all too often, examination of the facts by experts showed that there wasn't a mystery at all.

Even so there remains a handful of sea mysteries which continue to exert a fascination down to the present day, and which remain unsolved. Various explanations have been offered but none ever proved. This book investigates seven such mysteries, perhaps the most intriguing of all. Though ghost ships, monsters, mermaids and sea serpents are part of the legends that grew up around these seven cases, the actual circumstances did not produce anything either supernatural or connected with monsters.

These mysteries unfolded in different oceans approximately over the last 100 years, roughly 20 years apart. They are all different and relate to different types of vessels, from a giant liner to a multihulled racing yacht. In some cases the crew vanished, in some both the ship *and* the crew, but in almost every case the personality of a strong character is stamped across them—either to affect the course of events or in the way the mysteries developed.

Of the seven mysteries, one was solved fairly quickly and another after the passage of many years. With three more, a likely solution was proposed but was never proved. For the remaining two, even courts of inquiry were able to establish only that the identifiable events had occurred: beyond that there was only speculation.

1. H.M.S. Erebus and Terror (†1847)

Lost in the Arctic

For hundreds of years the frozen area north of the American continent dominated men's interest. What lay there? Was it possible to live there? What was it that caused compasses always to point in that direction?

The first attempt to learn something of this icebound area was made with three tiny craft in 1576 by Yorkshire-born Martin Frobisher. He discovered the great inlet now known as Frobisher Bay. More expeditions followed and gradually it was assumed that somewhere among the ice and frozen islands between Greenland and Russia was an open passage that led around the north of Canada to the Pacific—the Northwest Passage. From then on one expedition after another attempted to discover this almost legendary opening to the sea on the west side of the Americas. Among them was that led by Henry Hudson who, after visiting Spitzbergen, Novaya Zemlya, North America and the Canadian Arctic, was set adrift with a few companions in an open boat by a mutinous crew. The East India Company, interested in a route to its eastern possessions, also sent an expedition and, despite the failures, the idea that there was a northern passage to the Pacific persisted.

At the beginning of the nineteenth century, Britain was mistress of the seas and eager to test her power and extend her sovereignty. With Napoleon defeated, her desire for expansion took the form of exploration. In 1816 and 1817 the northern waters had been remarkably free from ice, making navigation easier and the prospects for Arctic exploration seem more promising than usual, so that once again explorers became interested in the Northwest Passage and the North Pole. The new era of Arctic exploration came to an end only in 1924, with the advent of polar aviation.

In 1817 interest in this long-talked-of route was revived largely because of Sir John Barrow, Secretary of the Admiralty, who sent out two expeditions in 1818, one under Lieutenant John Franklin, the other under Commander John Ross. More voyages were undertaken by Lieutenant Edward Parry, Captain George Back and Captain James Ross. All these men became distinguished explorers and all were subsequently knighted. They all left cairns with instructions and navigational records for the help of any who followed and, during their journeys by ship, boat and sledge, more and more information was added to existing knowledge of the northern coastlines.

By the 1840s all that remained to open the Northwest Passage seemed to be the discovery of a channel connecting Barrow Strait either directly with Bering Strait or with the waterway known to exist between Bering Strait and the west coast of Boothia. This need prompted the expedition of 1845, led by Sir John Franklin, who had first gone to the north in 1818.

The man behind the idea was again Sir John Barrow, who by this time was something of an institution at the Admiralty. Barrow had little time for anyone whose views ran counter to his own. He had written a book on the polar regions, believed himself to be an expert and was inclined to belittle such giants as Sir John Ross and Sir Edward Parry, claiming that they set too much store by the safety of their ships and men. "There can be no objection with regard to any apprehension of the loss of ships or men," he said. "Neither sickness nor death occurred in most of the voyages to the arctic regions north and south." He wasn't entirely accurate, but nobody noticed that.

For seven years there had been no government-inspired expedition to the north, despite the persistent urgings of the Royal

Geographical Society. Each year interest had grown as the Hudson's Bay Company's land expeditions carried out their coastal surveys, and at last it was felt that the lands and waters had been sufficiently charted for ships to sail from the Atlantic to the Pacific on known waterways. When it was suggested that H.M.S. *Erebus* and H.M.S. *Terror,* which had been used by Sir James Ross on three Antarctic voyages between 1839 and 1843, should be fitted with steam propulsion and sent off to try to reach the North Pole, Barrow proposed that instead the Northwest Passage should be their objective.

The proposals were examined by the Royal Society, Sir Edward Parry, Sir James Ross and Sir John Franklin. Significantly, neither Ross nor Parry liked Barrow's suggested course southwestward from Cape Walker, which they believed would lead into heavy ice, but they all approved of the idea, though Ross's uncle, Sir John Ross, prophesied that the expedition would disappear without trace.

Barrow had intended to give the command of the new expedition to a promising young commander in his early thirties, called James Fitzjames, but the Board of Admiralty believed that experience rather than youth was a more important quality in a leader now that the north had been opened. When the obvious choice, Sir James Ross, who had stood at the North Magnetic Pole and come within 160 miles of doing the same at the South Pole, turned down the command on the grounds of age (he was 43), it was offered to Sir John Franklin, a man 15 years older.

Franklin was more than eager. "Nothing," he said, "is dearer to my heart than . . . the accomplishment of the Northwest Passage." Parry, the only other major contender, said, "If you don't let him go, the man will die of disappointment."

Franklin was a man with a great reputation. He had been with Nelson at Copenhagen at the age of 15 and had distinguished himself, again with Nelson, as signal midshipman aboard *Bellerophon* at Trafalgar in 1805. He had been shipwrecked on an uncharted reef off Australia but had brought his survivors across 50 miles of open sea in a little boat to Sydney. In 1812, at the Battle of New Orleans, he had commanded a division of small ships against superior American gunboats. His uncle by marriage was Captain Matthew Flinders, famous for his discoveries in Australia, and Franklin had sailed with him when *Investigator* was sent to survey the coast of Australia.

By 1828 Franklin was considered to be one of the world's most experienced Arctic explorers. He had made his first expedition in 1817 and another two since then. Despite failing in his objective of finding the North Pole, he was considered to have distinguished himself enough as both an explorer and a leader to be promoted to commander after his second expedition, and he was later knighted by George IV. He was married a second time to Miss Jane Griffith, a beautiful, intelligent and stubborn woman of Huguenot descent. (His first wife had died after giving birth to a daughter, Eleanor.) In the 1830s he was given the governorship of Tasmania, an office which was made miserable, however, by the imperious attitude of the British government toward the colonists and by petty intrigues within the Colonial Office.

Franklin remained interested in polar exploration and, with only a few hundred miles left uncharted and the lure of the Northwest Passage remaining, he suggested to the government that one more well-organized voyage would clear up the problem. Because of his experience and reputation, he seemed a sound choice to lead the new expedition. He was recalled from Tasmania in 1843 and, with Ross's well-tried ships *Erebus* and *Terror* under his command, there was little fear of failure.

Save, that is, for the fact of Franklin's age. Parry had been only in his thirties when he went to the Arctic, but Franklin was 59 and beginning to stoop. "Actually 58," he said, when confronted by Lord Haddington, First Lord of the Admiralty, with this fact, "not 59 until April." Franklin was also said to suffer from the cold. A contemporary daguerreotype depicts him as a stout gentleman with a beam greater than the width of his shoulders. Nevertheless, he was much admired and very much liked by the men under him. According to Marine Sergeant David Bryant, in *Erebus,* he was "a pleasure to be with."

Command of the second ship was given to Captain Francis Rawdon Moira Crozier, who, though he regretted that Ross had not accepted the expedition's command, had great experience as a polar explorer, having served as a midshipman during Parry's second and third expeditions, as a lieutenant in Parry's fourth expedition and as captain of *Terror* during Sir James Ross's Antarctic expedition. He was 49. James Fitzjames, Sir John Barrow's choice as leader, who was only 33 and had distinguished himself in the China War, accepted the appointment as executive officer on Franklin's ship. He was a strong, self-reliant man, a good sailor and a born leader, and the men he had chosen were the cream of the navy. Commander Graham Gore had sailed with Back, and John Irving, Charles F. des Voeux and James Walter Fairholme had all done well in Africa and Australasia.

Fitzjames was also a great advocate of the new steam power that had been proposed, but in the end all they got were two small auxiliary engines and a screw to assist the sails and "propel the ships a few knots in calm." They were 20-horsepower machines, weighed 15 tons and were designed for the Greenwich Railway. Irving considered that they had done better as railway engines and would probably startle the Eskimos with their puffings and screamings. He didn't expect them to be used if there were other means of propulsion. Nevertheless, *Erebus* and *Terror* were the first screw steamers to be used in the Arctic and they had one notable advantage in that they carried hot water by pipe to the forecastles and cabins. Sanitation was a simple matter. The buckets were to be emptied overboard.

The officers took their own silver tableware. Among the items placed on board to keep them happy were a backgammon board and over 1000 books, including many devotional works, common in that day of high morality. There were also a dog, a monkey and an apparatus for making daguerreotypes. The stores, intended to last for three years, included preserved meat and vegetables, tobacco, 3000 gallons of overproof rum, warm clothing, wolf-skin blankets, 200 cylinders for messages which were to be thrown overboard in the hope that they would be picked up and sent to London, large stocks of lemon juice to combat scurvy and 10 live oxen, to be killed on the edge of the ice to provide fresh meat. By this time the Lords of the Admiralty were wholeheartedly behind Franklin and had provided liberally for the comforts of the explorers. As one of them said, "With the facilities of the screw propeller and other advantages of modern science, the expedition may be attended with great results." Though the equipment was woefully inadequate by modern standards, the men were going to sea in the most comfortable conditions then available, and in one of its outbursts of pomposity, *The Times* announced, "There appears to be one wish of the whole of the inhabitants of this country . . . that the enterprise . . . may be attended with success."

For their size, the ships were the strongest in the navy. Built as bomb vessels, they had frames and underworks in the forepart far stronger than in ordinary vessels to take the weight and recoil of heavy mortars. Before going to the Antarctic, they had also been provided with inner hulls so that their sides were dense masses of wood varying in thickness from 18 to 27 inches, while their bows, in addition to being sheathed in iron, were so reinforced as to contain eight feet of solid timber.

Erebus and *Terror* were not quite full sisters, the former being 370 tons against the other's 340. They had originally been ship-rigged, but Back, who had taken *Terror* to Hudson's Strait, had done away with the yards on the mizzenmast and *Erebus* subsequently underwent the same modifications. They were flush-decked, very broad of beam and had solid structures forward instead of the customary figureheads. For their auxiliary engines, however, only a meager 25 tons of fuel were carried. They carried a boat on either quarter, three on booms aft and two special whaleboats. Their provisions were intended for 137 people—though, because of vacancies and invalidings, only 129 entered the Arctic—and a supply ship, *Barretto Junior,* carrying a further supply of stores, was to accompany them to Davis Strait to top up their supplies before they entered the ice.

Franklin's orders instructed him to make his way through Lancaster Sound and Barrow Strait, which it was believed offered the best prospects of leading to the Northwest Passage, but not to examine any channels leading northward or southward from Barrow Strait, instead to push westward toward Cape Walker and then try to penetrate southward and westward toward the Bering Strait.

The ships left Woolwich for Greenhithe on May 12, and as they did so a dove settled on the mast of *Erebus,* which was considered an encouraging augury. Unfortunately there had been at least one of the other kind. Lady Franklin had stitched a Union Jack to be hoisted when the Northwest Passage was discovered and when, one cold day, she threw it over Franklin, he shuddered. "That's how they treat a corpse," he said.

At Greenhithe the crews were paid in advance so that they could provide for their families. The few who were not married kept the Greenhithe taverns busy. The crews were happy under Franklin. He was genuinely admired, and his officers already seemed to have settled down well. Crozier was a kind man of considerable experience. Des Voeux was "clever, light-hearted, obliging"; Gore a man of great stability and good temper, who played the flute with greater enthusiasm than skill. Harry D. S. Goodsir, the Curator of the Edinburgh Museum, who was accompanying the expedition, was 28 years old, tall and easygoing. Edward Couch was small, dark and quiet but of an excellent humor. Lieutenant H. T. D. Le Vesconte was easy to get on with and Fairholme smart and agreeable. These men were all described by Fitzjames in a journal he kept for his sister and, if nothing else, they seemed to have had the sort of temperaments that would bear up well after months of living on top of each other. In addition, morale, a highly important factor in the Arctic winters, was high.

The Lords of the Admiralty were among the crowd that watched the expedition up-anchor on May 19, 1845, and move upriver under tow, but Lady Franklin saw her husband once more and managed to wave to him at Harwich. They were held by stormy weather at Aldeburgh en route for the Orkneys but reached Whalefish Bay, near Disko in Greenland, in early July. Here they topped up from *Barretto Junior,* which was then sent home, taking with her one officer, two invalided men and *Terror's* sailmaker and armorer, who had been found to be useless at their trades. One man had earlier been sent home from Stromness.

Letters brought home by the supply ship, which suggested to friends that they should send their replies to Petropavlovsk, indicated that confidence was high. Fitzjames wrote to a school friend that they expected to get through the Northwest Passage that year and said that Franklin was highly respected. Of him, Fitzjames wrote, "He is in much better health than when we left England. He takes an active part in everything that goes on. Of all men he is most fitted for the command of an enterprise requiring sound sense and perseverance." Franklin himself made it clear in a letter to Sir James Ross that he felt he could find a route through the mass of islands which he believed occupied the blank space still remaining on the charts.

The ships set sail from Greenland on July 12 and were probably sighted at Upernavik. They reached the ice near Lancaster Sound, where they were met by two whalers, *Prince of Wales* and *Enterprise,* on July 26 in latitude 74 degrees 48 minutes N and longitude 66 degrees 13 minutes W. *Prince of Wales* was visited by a party of officers. Captain Martin, of *Enterprise,* spoke to Franklin and last saw the ships made fast to an iceberg on which they had set up an observatory. The camp was comfortable and even had its own newspaper. The men were in excellent spirits, though the death of two men from unknown causes had marred them a little. Franklin informed Martin that they had enough supplies for five years and his men were already busy salting down seabirds for additional food.

That was the last that was ever seen of Franklin or his ships. Franklin's, one of the most ambitious of all the Victorian expeditions, was to provide the most famous of all the Arctic tragedies.

When 1846 arrived few people in England were concerned about Franklin's expedition. The first Atlantic cable was not laid until 1858 and was not successful until 1866, and communications were poor. None of the 200 message cylinders had turned up, but still no one worried, though one veteran polar explorer, Dr.

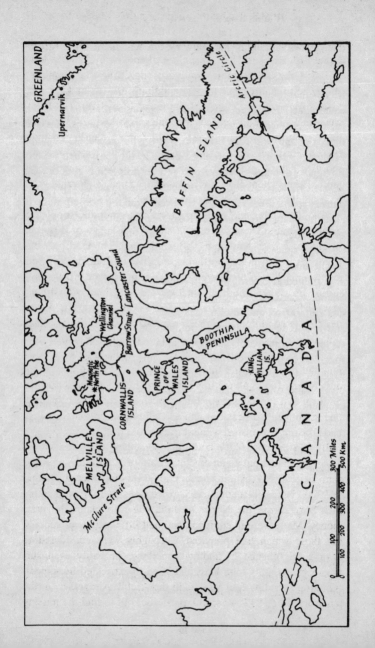

Richard King, who had accompanied Back on his trip down the Great Fish River in 1833, was alarmed enough by 1847 to write to Earl Grey, the Colonial Secretary, suggesting that Franklin's ships had probably come to grief and that survivors had better be looked for along the course of the Great Fish River. His letters were not even acknowledged. Sir James Ross had said there was no reason for anxiety and that Franklin's men would never under any circumstances have made for the Fish River, and other authorities agreed. Admiral F. W. Beechey, himself a polar explorer, after whom Beechey Island was named, alone thought a boat should be sent down that river. Nothing was done.

Another winter and summer passed and gradually people became anxious. There were still hopes that news of Franklin would come from Russia, but no word came. The Hudson's Bay Company combed the northern coastline of Canada and sent extra supplies to its northernmost stations in case the missing men turned up. When another winter passed and there was still no news, a feeling of bewilderment set in. Expeditions to the Arctic, despite the conditions, had been remarkably successful for some time. Hardships had been endured and individuals had died from a variety of causes, but by 1848 the Victorians, after their fashion, had begun to think that they had polar exploration well under control. It was all the more staggering, then, when it dawned on them that this latest expedition, under the command of someone as eminent as Franklin and so well organized and supplied, had apparently vanished without trace.

By now there was little doubt that something had gone wrong. The government put up an official reward of £20,000, a colossal sum by today's reckoning, and Lady Franklin offered another £5000. Naturally enough, 15 expeditions were soon making their way across the North Atlantic eager to claim the prize money. Many were ill equipped and had little chance of success, but those which had prepared thoroughly were considered to have a fair chance of finding something, probably even survivors, because Sir John Ross had after all extricated his expedition after spending four winters in the Arctic. Altogether, in the ten years after 1847, 40 search parties set out to find the missing explorers, six of them going overland to the coast of Arctic America, and 34 exploring the waterways among the maze of northern islands. Some of the searchers proved themselves to be explorers, too, and though at first their task was solely to find survivors or some trace of them, the journeys they made filled in many of the blanks on the map.

When nothing had been heard by 1848, Sir John Richardson was sent to examine the coast between the Mackenzie and Coppermine Rivers but not to go to the Fish River, and Sir James Ross, a personal friend of Franklin's, was sent to try to find some trace of the missing expedition. He went, however, in the full conviction that he would meet the missing ships returning home or would return to find them already in the Thames. Nevertheless, sledging parties were sent out, provision dumps set up and several men died, but nothing was found. When Ross was forced to return home as his supplies ran out, the Admiralty decided that Franklin could not have gone south along the west coast of North Somerset. However, if nothing else had been achieved, advances had been made in the technique of sledging. With Ross had been a young lieutenant, Leopold McClintock, whose short, wiry frame had proved ideal for long exertion and hardship, and, though threatened by scurvy, he had managed to note down every detail about the sledge parties—equipment, food, weight of the rations, clothes and construction of the sledges—so that they could be pondered over with a view to improvement.

It was not until 1850 that an official squadron under the command of Captain Horatio Austin found the remains of Franklin's first camp. There were coalbags, rope, cinders, the remains of a small garden and a large number of opened tins filled with gravel, which gave rise to the belief that much of the expedition's preserved meat had been unfit to eat. In fact the alarm was groundless as many of the tins were marked "soup." There were also traces of a carpenter's and an armorer's working places, washing tubs, old clothing and the graves of three men, Leading Stoker John Torrington, of *Terror,* and Able Seaman Thomas Hartnell and Marine William Braine of *Erebus,* who had died between

January and April 1846. There were also sledge tracks, one set heading into the interior.

These were the first signs of the lost expedition. Then the Eskimo interpreter who had been with Ross came up with a story of men with epaulettes having been killed by Eskimos with darts and spears, a story he had gleaned from local men. The story also told of two ships, each with three masts, near Cape York in 1846. Another Eskimo story was that white men had bartered arms for food, but had died and been buried near the Mackenzie River, and there was a further account of murdered white men. But it was felt that the stories were not necessarily true but were attempts by Eskimos to please people who were desperately seeking information.

However, one of Austin's captains, Robert le Mesurier McClure, of *Investigator,* standing 600 feet above sea level, looked northward and, seeing only sea, realized that he was looking at the Northwest Passage. Ironically, the men searching for Franklin had come on the object of Franklin's search. The following day McClure shot a large bear in the stomach of which were things indicating the proximity of civilized men—raisins, a few pieces of tobacco leaf, bits of pork fat cut into cubes and surgeon's sticking plaster. They felt they were on the trail of Franklin at last. Lieutenant Samuel Gurney Creswell was sent back with dispatches indicating what had been found.

But no more traces came to light until May 8, 1851, when startling news of Franklin came from an entirely different source. *Erebus* and *Terror* were believed to have been sighted by the two-masted brig *Renovation,* which had left Limerick, Ireland, on her way from North Shields to Quebec. The *Limerick Chronicle* published a letter from the uncle of a Mr. Lynch, who had been a passenger in *Renovation,* describing icebergs the ship had met, of such a size that alongside them the steeple of a cathedral would have appeared a small pinnacle. One of them, it was reported, had had "two ships on it, which, I am almost sure, belonged to Sir John Franklin's exploring squadron."

At first nobody took much notice. In those days the Irish were considered a joke by English papers and Irishisms abounded in *Punch*. But the master of a German ship, *Doctor Kneip,* reported having seen later the same month two waterlogged ships which sank soon afterwards, close to where the iceberg had been seen by *Renovation*. It sounded very much as though they were the same ships and that the iceberg had broken up. Still no one took much notice, until suddenly, on March 2, 1852, nearly a year later, everybody woke up to what had happened and wanted to know more.

The key witnesses, it seemed, were Robert Simpson, the brig's mate, and Joseph Lynch, the passenger. On about April 18 or 20, the previous year, Simpson said, they had fallen in with numerous icebergs near the Newfoundland Banks. While on the morning watch he had seen two vessels stranded on an iceberg.

> One was lying on her beam ends, with her deck towards us, having only her lower masts and bowsprit left standing. The other had her topmast on end, with lower and topsail yards across, but no sails bent; she had no topgallant masts up, and was nearly end-on to us. She was lying on an elevated part of the ice, far above the other one. I went down and called the Master, who was lying very ill in bed, and reported the circumstance to him. At first he didn't speak; I mentioned it a second time and then he said, "Never mind." I returned on deck and stood on the lee quarter to watch the vessels; all the watch came aft to look at them also, with the spyglass. I called Mr. Lynch (a passenger) out of his bed, who came immediately in his shirt and drawers, but afterwards went down and dressed. . . .

Lynch thought they might be Franklin's ships as they were so close together. The mate disagreed. "I regarded them as wrecks," he said, "and therefore did not trouble my mind further about them at the moment."

There were no signs of life on the ships and nothing could be discerned but the hulls, masts and yards. Since the ships were about three miles away it was only just possible to distinguish these details with the naked eye.

The ships were kept in full view for three quarters of an hour. According to a sketch later drawn by Simpson, the ship high on the ice had had topmasts rigged and looked in perfect order; the other, with lower masts only, was heeled over to starboard, so near to the edge of the berg her yards seemed to overhang the sea. No boats were visible either aboard or on the ice. The ships were broad in proportion to their length, as were *Erebus* and *Terror;* they were flush-decked, which was an almost sure sign they were not whalers, which had raised forecastles; their hulls were black and masts light, like Franklin's ships; and their bottoms did not appear to be coppered, another similarity. The only noticeable difference was that the upright ship was said to be ship-rigged, though as she was seen in a nearly end-on position at a distance of three miles, this description could well have been none too reliable.

It was suggested that the episode arose out of a hoax played on Lynch, a raw Irishman, but in fact Lynch was a well-educated man who had himself followed the sea for four years. Simpson became the principal witness at the subsequent inquiry and Lynch gave corroborative evidence. The sighting was also reported independently by a crew member in a letter. If Captain Coward, the master of *Renovation,* had not been so ill and some attempt had been made to move closer, much more might have been discovered.

In 1852 another naval squadron, *Assistance, Resolute* and *North Star* (sail) and *Pioneer* and *Intrepid* (steam), left to search Wellington Channel and Melville Island, but, having passed over all the available outstanding men, the Admiralty offered the command of the squadron to Captain Sir Edward Belcher. He was "an old officer, with bad health, no Arctic experience and the reputation of being the most unpopular man in the navy." Indeed, Fitzjames had once said that he would go anywhere in any capacity, but not as second to Belcher.

Even the *Dictionary of National Biography* has little good to say of Belcher. "The appointment was an unfortunate one," it observes dryly. Though able and experienced, Belcher had neither the temper nor the tact necessary for a commanding officer under the circumstances of peculiar difficulty that existed. "Perhaps," it continues, "no officer of equal capability has ever succeeded in inspiring so much personal dislike." His expedition was a failure and it was "distinguished from all other Arctic expeditions as the one in which the commanding officer showed an undue haste to abandon his ships when in difficulties, and in which one of the ships so abandoned rescued herself from the ice and was later picked up floating freely in the Atlantic."

With the Admiralty directing Belcher toward the upper portion of the Wellington Strait, because they were convinced that no further searching to the south of Cape Walker was necessary, the expedition was a complete fiasco. Not a trace of Franklin was found and, for no good reason, Belcher abandoned four of his five ships when they were caught in the ice and squeezed the crews into the fifth. He was court-martialed on his return and, though acquitted, he was damned forever as a bungler. Fortunately, among Belcher's commanders were sound men such as Sherard Osborn and McClintock, who both added to their Arctic experience.

The disaster did not stop Belcher's publishing a book on the expedition in 1855, with the somewhat extravagant title *The Last of the Arctic Voyages.* He was never employed again, though in those days of the sanctity of seniority, he attained his flag in 1861 and rose to full admiral in 1872. He also published *Horatio Howard Benton, A Naval Novel,* described by the *Dictionary of National Biography* as "an exceedingly stupid one."

After the Belcher expedition the government refused to have anything more to do with Franklin. A few naval experts agreed with Lady Franklin, however, and urged further action. One letter to the government suggested that the investigations should be continued "to satisfy the honour of our country and to clear up a mystery that has excited the sympathy of the civilized world."

Meanwhile, every crackpot amateur scientist imaginable and every kind of medium—one of them a child—offered theories. One man even hired a London hall in 1854 to announce that, since compass needles indicated large quantities of electricity were continually traveling toward the pole where it condensed and caused combustion, Franklin might well be in a genial climate but, without fuel and steam, unable to return, "on account of the constant wind rushing towards the Pole to feed the fire."

Lady Franklin, who had never given up hope, had by this time spent a great deal of money in funding the search parties, even spending part of a legacy which should have gone to Franklin's daughter, Eleanor, now married to an impecunious clergyman, Philip Gell. Eleanor had given up all hope for her father after seven years. As Franklin's will had settled his money and property on her, leaving Lady Franklin only a life income and no capital, that redoubtable lady now approached her father. He realized that anything he left his daughter would be spent in further search for her missing husband, and so when the old man died in 1852 he was found to have cut her out of his will in favor of his grandson, with whom she was not on good terms and from whom she could expect no help.

Lady Franklin continued to spend, however. She had sent the ship *Prince Albert* off in 1850, with orders to head southward along Prince Regent Inlet. Though she was not exact, she was right in her assumption that the solution to the mystery lay to the south of Lancaster Sound. *Prince Albert* found nothing, so she sent it out again and even chartered another ship, *Isobel*, before she knew that *Prince Albert*'s search had been unsuccessful.

But it was her wish to set up a memorial to a young French naval officer who had died in one of the ships she had financed that proved the last straw for the Gells. A family feud that had started over her spending broke out afresh. The Gells insisted that Eleanor's money should be spent by Eleanor, not by Lady Franklin, and letters appeared in *The Times* from the supporters of both sides. But when the Admiralty announced that it was about to remove Franklin's name from the Navy List, only Lady Franklin's voice was raised in protest. Finally, on the presumption that her father was dead, Eleanor got her money.

Then late in 1854, with the mystery apparently closed, Dr. John Rae, Chief Factor of the Hudson's Bay Company, while surveying the west coast of Boothia, came across an Eskimo who told him of white men who had starved to death on an island. He managed to piece together a story of about 40 men dragging a boat and sledges over the ice after their ships had been beset. Later, he learned, the graves and the bodies of about 30 white men had been discovered on the mainland and five more bodies on an island, some in tents, some outside, others under an up-turned boat. One had a telescope and a double-barreled gun. Cannibalism was suggested.

From his Eskimo informants, Rae also obtained the first evidence that Franklin had actually been in the Arctic—his badge of the Order of Hanover, a silver bowl with his crest, 20 silver spoons and forks with the initials of officers of the two ships, among them Gore, Le Vesconte, Fairholme, Couch and Goodsir, many smashed guns, watches, compasses, telescopes and other articles, including a waistcoat marked with the initials of Des Voeux, mate of *Erebus,* all of which, the Eskimos said, had come from the upturned boat.

Unfortunately, by the time Rae's report reached England, the Crimean War had started, all naval resources were fully occupied and the fate of Franklin's party seemed unimportant against the men dying in south Russia. Nevertheless, the Admiralty requested the Hudson's Bay Company to investigate further. In 1855 Chief Factor James Anderson turned up oars, tools and other objects, including a plank bearing the name *Terror* and even the back-gammon board Lady Franklin had supplied, but no sign of graves or bodies. Anderson decided from the discovery of wood shavings that a boat had been cut up by Eskimos, and it required Captain Sherard Osborn to point out that the shavings must have come from a plane, a tool unknown to Eskimos, and therefore were proof that white men had been there. Osborn's comments went unnoticed and Dr. Rae was paid £10,000 of the reward money.

By this time finding the case a nuisance, the Admiralty considered that Rae's discoveries effectively closed it. Lady Franklin, however, singleminded as ever, pointed out that, on the contrary, Rae's finds had not closed the case at all, but opened it. The discoveries, she insisted, proved only that *Erebus* and *Terror* had had to be abandoned, perhaps, but no more. "In fact," she said, "it does not even prove that." The men the Eskimos had seen might well have been a reconnaissance party, sent from the ships, who had lost their way. After all, only 35 bodies had been found by the Eskimos but the total party consisted of 129 officers and men. Were they to believe there was not a single survivor left? "The bones of the dead," she insisted, "should be sought for . . . their buried records . . . unearthed . . . and . . . their last written words . . . saved from destruction." She was met with a cold refusal.

By this time not only Lady Franklin's family but also everybody who had been to the Arctic and many who had not were quarreling about where to search, even whether to search at all. The matter was raised in the House of Commons and *The Times* thundered that, while Lady Franklin could do what she liked with her own money, it was far from right to spend public money for "so preposterous a scheme as another search of Sir John Franklin's relics."

Yet she never gave up and, organizing yet another new search, this time she gave the leadership to McClintock, who was fast emerging as the navy's greatest expert on the Arctic. Her friends put money into the expedition and more was raised by public subscription, the general feeling being that the government was not doing enough and had abandoned Franklin too soon. Perhaps to ease its conscience, the government did supply arms, shot, powder for blasting, rockets, maroons and signal mortars, 6628 pounds of pemmican, meteorological and nautical instruments and journals, ice saws, ice anchors and ice claws, together with books and clothing. The whole expedition cost £6000.

McClintock was given the 177-ton yacht *Fox* and a completely free hand. He had only 25 men, but all were young and all experienced. He sailed from Aberdeen on July 1, 1857, and 70 miles west of Upernavik in Greenland reached the edge of the "middle ice." He steered north and anchored off Beechey Island, where, clearly not expecting to find any survivors, he set up a memorial to Franklin and his men. Arriving off Melville Bay on August 12, the ship was beset and carried about 100 miles with the drift, her creaking timbers constantly subjected to the enormous pressures of the ice. In his diary McClintock wrote, "I can understand how a man's hair can turn grey in a few hours."

McClintock remained optimistic, however, setting down his thoughts in his journal, organizing a school and getting his men to play rounders on the ice, even holding a fireworks party on November 5, when they ate one of Lady Franklin's preserved plum puddings. The ship was freed by gales on March 27, 242 days after being beset and almost exactly 13 years after Franklin had sailed. McClintock promptly turned her bows north again and this time he reached Lancaster Sound. By August he had reached Beechey Island, where Franklin's opened tins and the Eskimos with the silverware had been found a few years before. Finally beaten by the Arctic winter, he decided to wait at the entrance of Bellot Strait, but, not wishing to waste time and despite the temperature of 60 degrees below zero, he began to explore the territory around.

On March 1, 1859, a party of Eskimos appeared, and one of McClintock's men caught sight of a British naval button glinting on the coat of one of them. Questioned by sign language, the Eskimos produced trinkets which included six silver spoons and forks and a medal bearing the name of Dr. Alexander McDonald, assistant surgeon of *Terror,* part of a gold chain, more buttons and knives, and bows and arrows which they said they had made out of the wreck of a ship. They knew nothing of Franklin and had obtained the relics by barter with other Eskimos, who told them of a big ship crushed by ice off King William Island. Feeling he was at last near to solving the 14-year-old mystery, McClintock planned an extensive sledging expedition. While he himself struck out southeast for King William Island· in an attempt to locate the ship the Eskimos had mentioned or the Eskimos who had originally found the relics, Captain Allen Young, a close friend of Lady Franklin's who had volunteered for the trip, was sent to investigate Prince of Wales Island where a second ship was said to have been seen. After trekking across the ice, McClintock split his party into two groups at King William Island. One, led by Lieutenant William R. Hobson, headed out along the west shore, while the other, under McClintock, set out southward with the idea of meeting on the other side of the island. If clues were found, they were to probe farther inland. Almost at once they became aware that they were in the right vicinity and the story finally began to unfold.

Hobson found a cairn obviously left by the expedition. It contained a blanket, collapsed tents, a sextant and some clothes, still well preserved, one article marked "FDV, 1845," which seemed to be part of the clothing of Charles F. Des Voeux, the mate of *Erebus*. Digging a trench round the cairn, the party then discovered broken bottles and a boat's ensign. A few days later and a few miles farther on at Point Victory, Hobson came across another cairn where, alongside the rocks, was one of Franklin's tin canisters containing a message:

28 of May [the message read], HM Ships, *Erebus* and *Terror,* wintered in the ice in Lat. 70 degrees 05 minutes N, Long 98 degrees 23 minutes W., having wintered in 1846–7 [surely an error for 1845–6] at Beechey Island in Latitude 74 deg 42 min 28 sec N, long 91 deg, 39 min 15 sec W, after having ascended Wellington Channel to lat 77 and returned by the west side of Cornwallis Island.

Sir John Franklin commanding the Expedition.

All well.

Party consisted of two officers and six men, left the ships on Monday 24th May, 1847.

The message was signed "Gm. Gore, Lieut.," and Chas. F. Des Voeux, Mate."

There was a clear note of hope and cheerfulness about the message, but it was immediately dispelled by smaller writing which filled the margins of the paper.

25 April, 1848 [the second message read], HM Ships *Terror* and *Erebus* were deserted on 22 April, five leagues NNW of this, having been beset since 12th September, 1846. The officers and crews, consisting of 105 souls, under the command of Captain F.R.M. Crozier, landed here in Lat 69 deg 37 mins 42 secs N, long 98-41 W. Sir John Franklin died

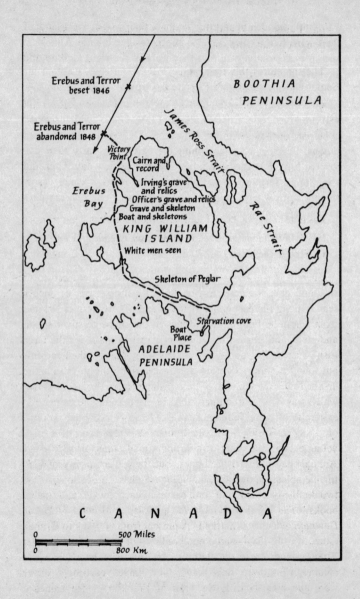

Erebus and Terror
beset 1846

BOOTHIA
PENINSULA

Erebus and Terror
abandoned 1848

James Ross Strait

Victory Point

Cairn and
record

Erebus
Bay

Irving's grave
and relics
Officer's grave and relics
Grave and skeleton
Boat and skeletons

Rae Strait

KING WILLIAM
ISLAND

White men seen

Skeleton of Peglar

Starvation cove

Boat
Place

ADELAIDE
PENINSULA

C A N A D A

0   500 Miles
0   800 Km

on 11 June, 1847, and the total loss by deaths in the expedition to this date, nine officers and 15 men.

The signatures this time were "F.R.M. Crozier, Captain and Senior Officer, and James Fitzjames, Captain, H.M.S. Erebus."

A postcript added, "And start tomorrow, 26th., for Back's Fish River."

There was one other message:

This paper was found by Lieutenant Irving under the cairn supposed to have been built by Sir James Ross in 1831, 4 miles to the northward, where it had been deposited by the late Commander Gore in June, 1847. Sir James Ross' pillar has not, however, been found and the paper has been transferred to this position, which is that in which Sir James Ross' pillar was erected.

There it was at last—real information! Hobson returned to the ship and gave his news to McClintock, who set out at once to search along the east coast of the island. There he met Eskimos from whom he bought more relics in the shape of silver spoons and plate, and learned that the survivors had made for the Great Fish River, falling and dying as they went. He searched carefully but found nothing. Turning northward along the western shore, first he found the skeleton of Petty Officer Harry Peglar, of *Terror*, which was lying in the open, then at Erebus Bay at the western extremity of King William Island, a large ruined boat mounted on a sledge. Inside were two human skeletons, one of a slight young person, the other of a strongly made man of middle age. Animals had ravaged the bones, but there were plenty of identifiable objects in the boat, including watches, guns, slippers, furs, two double-barreled guns, one barrel in each cocked and loaded, books—*The Vicar of Wakefield,* a Bible, *A Manual of Private Devotion,* which was marked "A present from G. Back to Graham Gore, May 1845,"—and a notebook full of indecipherable notes. There was also a lot of clothing, together with boots, handkerchiefs, towels, soap, nails, saws, files, knives, cartridges, silverware and plate bearing the crest of Franklin or the names of

Gore, Le Vesconte, Fairholme, Couch and Goodsir. Others belonged to Crozier, Hornby and Thomas. There were also 40 pounds of chocolate, tea and tobacco, but no biscuits or meat. McClintock came to the conclusion that the boat had been returning to the ships and had had to be left behind, and with it the two men who had been too weak to continue.

Fox arrived at Portsmouth on September 21, 1859, with two cases of relics. Franklin's fate at last seemed to be solved and for Lady Franklin the search was finally over, though it was still not known what Franklin had achieved, only what he had failed to do. There had been little hope for a long time, but now the full tragedy was clear. The only consolation to the relatives was that lack of food and exposure to cold produces a condition of apathy, stupor and insensitiveness to pain, so that the men must have died peacefully and without suffering.

Inevitably, the mood changed and, in view of the discoveries, it was now suggested that Franklin's men, instead of being the supermen everyone had imagined, were timorous and had clung to their ships too much. McClintock, with few resources and fewer men, had made a circuit of King William Island and had even found the only feasible route to the west. What was more, since both *Fox*'s engineer and his assistant had died—neither of Arctic causes—McClintock had even had to get the ship's engines going and handle them himself until she was clear enough of the ice to raise sail.

During the 1860s an American, Captain Charles F. Hall, of Rochester, New Hampshire, spent several years living with the Eskimos. Hall had been a blacksmith, a shopkeeper and a journalist, and he believed he had a good chance of finding out what had happened to Franklin, because he would be searching in the summer whereas McClintock had searched during the cold season. His offer to the American Geographical Society to equip an expedition was accepted, and with help from public subscription he fitted out the whaler *George Henry* and took with him a store ship, *Rescue*, a steam-driven paddle vessel, which had already been out once in 1850 on the search. On her first trip *Rescue* had been considered a bad-luck ship. The mortality rate among her crew had been high, and her sister ship *Advance* had been lost in the ice.

Hall's expedition was considered harebrained, but if nothing else he brought back a weird story that smacked of the supernatural. He reached Frobisher Bay on Baffin Island in September 1860. While the two ships lay at anchor a gale blew up which raged throughout the night of September 27. Living up to her reputation for ill-luck, *Rescue* went aground and began to disintegrate. Hall left, having heard rumors of white men living with Eskimos, and when in July 1861 he returned to Frobisher Bay, he looked for the remains of *Rescue* but found nothing beyond small fragments of wood.

A few days later, however, on July 27, *George Henry* was near Whale Island in the Hudson Strait when Hall was called on deck. About two miles away a battered hull was seen floating, and as they approached they realized it was *Rescue*. Curiosity gave way to fear when it was noticed that *Rescue* seemed to be steering some sort of course and the superstitious sailors began to talk of a ghost crew. Hall's first thoughts were that survivors of Franklin's expedition had come across the wreck and fitted her out with the help of the Eskimos. But there was no sign of life on board and that night, as *George Henry* was threatened by ice, it was noticed that *Rescue* was moving steadily toward her old consort. Something like panic set in because the hulk, which appeared to be under some sort of control, seemed set on ramming *George Henry*. Hall was even called on to abandon ship, but at the last moment *Rescue* slipped away astern and vanished. For two more days the two ships played cat and mouse while Hall, never daunted, continued to search the coast for the men he hoped to find. Not until August 5 did *Rescue* finally disappear seaward.

In addition to this strange story, Hall also brought back a mahogany barometer case, a silver watch, two spoons bearing Crozier's initials and the skeleton of Lieutenant H. T. D. Le Vesconte, of *Erebus*, whose grave he had found. He also brought information of four men, one of them believed to have been Crozier, who had survived the general disaster and lived for a while with the Eskimos.

More relics were found in 1878-79, by a United States Army officer, Lieutenant F. Schwatka. He led an expedition backed by the American Geographical Society which found the grave of Lieutenant Irving, of *Terror*, whose skeleton was sent back for burial in Scotland. During the present century other skeletons, bones and relics have been discovered. But since McClintock's discoveries of 1859, no further light has been shed on the disaster. Neither Franklin's grave nor any more records have been located. Just how awful conditions must have been was discovered by an expedition in the bitter winter of 1875-76, yet sledge parties, led by Commander Albert Hastings Markham, had managed to drag not only their laden sledges but also two boats through belt after belt of hummocked ice.

Judging by the cheerful "All well" message left by Gore, conditions were not bad in May 1847, however, and in 1866 Captain Hall met an Eskimo who said he had seen Franklin in the early months of that year, lame, sick, but still cheerful and optimistic. McClintock always believed that Franklin's men, if not Franklin himself, arrived at the missing section of the Northwest Passage but had not lived to tell the tale.

The story, as it was pieced together, seemed to be that *Erebus* and *Terror* had successfully penetrated the Barrow Strait but, finding ice conditions unfavorable, turned into the Wellington Channel and moved to a relatively high latitude of 77 degrees N. At that point they found their way blocked and returned to Beechey Island, where the explorers found plenty of game. The second summer they left Beechey Island, found the inlet of Franklin Channel and went down it between North Somerset and Prince of Wales Island until they reached open sea. Had they gone between King William Island and the coast of Boothia they might have been successful, but, believing that King William Island was a peninsula, they passed outside it into the spiderweb of the Arctic and they became beset by ice a short distance from the northern end of the island on September 24, 1846. According to Eskimos, the following summer was the coldest in living memory and the ships failed to extricate themselves. They spent a third winter drifting with the ice. The following spring Franklin sent out Gore with a party to find two cairns which should have contained vital navigational information, one left by the Hudson's Bay Company, the other by Sir James Ross 15 years earlier. Only 150 miles away, unknown to either, Rae was mapping the last of the uncharted east coast.

Apparently Gore returned, having failed to locate the cairns but having left the first message, to find that while he had been away Franklin had died on June 11, just after his 61st birthday. No record was found of the cause of his death, and while Franklin's memorial suggests that his body had been consigned to the sea through a hole cut in the ice, there are reasons to believe that it was buried among rocks on King William Island and concreted in with cement carried for the bedding of the steam engines. It is possible that the records of the expedition were buried with him, and that the grave was later broken open by Eskimos and Franklin's bones and the records scattered.

Crozier had now taken over command, and when the partial thaw came in July, he and his men found the ships had been carried about 20 miles southward by the ice. When winter came again, locking them in once more near Victory Point, the men began to die, Gore, eight other officers and 15 men succumbing first, probably of exposure, frostbite and exhaustion. By this time the ships had become a burden and Crozier made the calamitous decision that the only hope of survival was to abandon them and march over the ice to King William Island and cross to a point from which they could reach the mainland. From there they stood a chance of making their way southward to the Great Fish River, where help could be obtained.

On April 22, 1848, the 105 survivors abandoned the ships and started to drag the ships' boats laden with their supplies across the ice. They were starving and faced a murderous 250-mile journey. Against blizzards and extreme temperatures, they grew weaker and in three days managed to cover only fifteen miles. At Victory Point they added a second message to Gore's earlier note, writing in the margin, and, deciding speed was essential, Crozier now abandoned some of the boats and provisions so that they carried only their barest needs. On April 25 he left the cairn with the message which ended, "And start tomorrow, 26th, for Back's Fish River."

At some point they found the trek beyond them and decided to turn back to the ships and await a possible thaw. At that time there must have been more than 90 of them still alive. According to Eskimo stories, four men survived until 1849, but what happened to them is a mystery. Not one man of the 129 who went into the Arctic reached civilization.

According to Sir John Richardson, by their march they had forged the last link of the Northwest Passage, and Sir Clements Markham considered their fate was made more melancholy by the fact that some of them might have been saved. When no news arrived in 1846, he believed, prompt measures should have been taken. But the Admiralty did nothing. Even in 1848 Sir John Richardson might have saved a few, had his instructions allowed him to search the Fish River area.

The Admiralty's decision that Franklin could not have gone south proved to be a grave error and the tragedy now seemed needless, because Dr. King, in his letters in 1847, had even pointed out the spot where the ships had been beset. Their fate might well have been discovered by *Prince Albert* on her second voyage, had the captain obeyed his instructions. Dr. Rae, in 1854, might even have seen them still caught in the ice and rescued any records that were aboard the ships had he been able to see across the 50 miles of Victoria Strait.

It was not until 1904 that Roald Amundsen, the Norwegian, following the route McClintock had discovered, found that there was in fact a passage through to the west. The great hopes that had rested on it never came to anything, however, because it was always too difficult, and in 1920 the Panama Canal made a northern route to the Pacific unnecessary. Nevertheless, Lady Franklin had been vindicated. In May 1860 the Royal Geographical Society's Founders' Gold Medal was presented to her for her persistence and she was reconciled with her stepdaughter.

With Franklin's fate clear, polar exploration did not end. Nor did the tragedies. In July 1879 Lieutenant George Washington

De Long of the U.S. Navy sailed from San Francisco for the Pole and when his ship was beset, out of 32 men, only two, who were sent off for help, survived. Even as De Long died in 1881 another American expedition, this time under Lieutenant Adolphus W. Greely, of the U.S. Army, was setting off. Twenty-six men were landed high on Ellesmere Island but, though they collected an immense amount of data, the relief ship which should have arrived in 1882 did not appear and another attempt in 1883 failed. When the survivors were rescued in 1884 by Captain W. S. Schley, only six of them were alive.

Franklin is as real today to explorers as he was 100 years ago. During the search for the lost Italian airship *Italia*, which crashed on polar ice in 1928, human remains and artifacts were found, but these were associated not with airships but with wooden sailing ships and were identified as coming from survivors of the Franklin expedition. They seemed to indicate murder and cannibalism as several of the skulls had neat shot holes in them. Nowadays land parties of Canadians and Americans working north of Hudson's Bay to service the Distant Early Warning radar line, a keystone in North America's air defense system, still half expect to come across Franklin's remains.

In 1978 Roderic Owen, a descendant of Franklin, went over the whole sad story of the Franklin expedition again. He could only be certain that the movements of the crews after the ships were beset by the ice were merely those of men fighting for survival. In 1976 Owen himself had gone to Alaska to unveil a plaque commemorating the 150th anniversary of Franklin's discovery of Prudhoe Bay, now the biggest oil field in North America. Leaving England before breakfast, with a flight over the Pole and with the advantage of the International Time Scale, he arrived at Anchorage in time for coffee, then flew up to Prudhoe and, using the Franklin expedition's own map, completed in less than one day a journey that had taken Franklin in 1826 a year and a half.

And the fate of the ships? Various views have been taken. After they had been abandoned they must have drifted down Victoria Strait in the ice, and Eskimos told McClintock that one of them had sunk somewhere to the west of King William Island. Later discoveries suggested that she passed through Simpson Strait and sank on the east side. The other ship went on down Victoria Strait and then toward Simpson Strait, where she was said to have been found by Eskimos in 1849, literally at the entrance to the last link in a Northwest Passage. How much strength can be attached to the Eskimos' report, however, is dubious. They described the ship as carrying five boats and being housed over with awnings, and having a gangway leading to the ice. Their statements suggested also that she had been inhabited during the previous winter and that though no living person was found aboard, the frozen body of a lone white man was discovered. Later the ship sank, leaving only the topmasts visible, and footprints were found on nearby land, probably made by the last survivors of the expedition, one of whom had died in the ship, though Rear Admiral Noël Wright, an ardent inquirer into Franklin's fate, suggested that the body that was found could well, from its description, have been a figurehead which had become a human being in the dubious translations.

A curious story appeared at the beginning of the twentieth century to the effect that the masts had been seen by an Indian with Chief Factor James Anderson's search party in 1855, but, realizing that an attempt to reach the ship that supported them would be dangerous and would delay the expedition's return home, he did not report them so that Anderson never knew about them. Another later Eskimo story was that a ship could be seen lying under the water in James Ross Strait, which could well have been the same ship. Unfortunately the Eskimos, knowing that the searchers wanted news of the lost ships, were often only too willing to provide it from their imagination, and their idea of time was so vague that no dates could be fixed, while their reports were always confused.

Their description of the awnings over the ship made sense, however, because it was the practice to use awnings to keep the decks snow- and ice-free and to provide extra warmth and shelter. But the reported presence of five boats seems strange because several boats or the remains of boats were found, which seems to indicate that some at least of the boats would be missing. It would be only too easy for an Eskimo, asked in an attempt to identify a ship if there were five boats on board, to nod and smile and agree.

As for the ships seen on the iceberg by *Renovation,* their description seemed to tally very well with that of *Erebus* and *Terror*—black hulls, white masts, the fact that they seemed to be consorts, one slightly larger than the other. They had apparently been carefully dismantled and the yards stripped, which would make sense with men trying to salvage everything possible that might add to their safety. They also bore a striking resemblance to *Erebus* and *Terror,* but it is impossible to say whether the two waterlogged derelicts sighted the same month near Newfoundland by the master of the German ship *Doctor Kneip* were the same vessels.

According to Rear Admiral Wright, there is little doubt that the ships *were Erebus* and *Terror* and, though other ships had been lost in the attempt to find Franklin, there had been no case of two ships' being lost together. The ships on the iceberg certainly did not look like whalers, and no losses which could not be accounted for had occurred during the previous three years.

But how did they reach Barrow Strait when the ice would normally have carried them in the opposite direction? It is just possible that a counter-drift could be caused occasionally by ice coming from McClintock Channel. They could have reached Barrow Strait and Davis Strait and from there moved into the Atlantic, because the easterly drift also carried with it H.M.S. *Enterprise,* H.M.S. *Investigator* (abandoned in 1854 and found almost intact in 1910), two American-equipped ships and Belcher's abandoned *Resolute,* which was reported in Arctic waters for four consecutive years before finally clearing the ice without assistance and turning up in 1907—*off Cape Horn!* The ship found by the Eskimos might well have been another of Belcher's ill-fated squadron, but, of course, the ships on the iceberg might equally well have been foreign ships, of which no report had been received. The evidence either way is too scanty to be able to offer an opinion.

Frank Worsley, who was with Sir Ernest Shackleton on the 800-mile open-boat journey that brought his expedition back to safety in 1915 after his ship, *Endurance*, had been destroyed by the Antarctic ice, described how, when they went ashore on South Georgia between Cape Horn and the Antarctic, they found themselves in a strange cemetery of ships. He described a pile of driftwood, covering half an acre and from four to eight feet high in places: ". . . lower masts, topmasts, a great mainyard, ships' timbers, bones of brave ships and bones of brave men. Most of it had drifted a thousand miles from Cape Horn, some of it two thousand miles or more."

Swept before the westerly gales onto the wild South Georgian coast, the easterly current, by some strange freak of eddies, had thrown it up in this one spot. Piled in utter confusion lay beautifully carved figureheads, well-turned teak stanchions with brass caps, handrails clothed in canvas coachwhipping and finished with Turks' heads, cabin doors, broken skylights, teak scuttles, binnacle stands, boats' skids, gratings, headboards, barricoes, oars and harness casks. Nothing, he said, was identifiable because the weather had eroded paint and worn away names, but they were all from ships which must have been reported lost without trace. Could it be that the bones of *Erebus* and *Terror* lay among them? If *Resolute*, lost in Arctic waters, could turn up near Cape Horn, surely so could *Erebus* and *Terror*.

Or could they still be in the Arctic? Early in 1979 came the news that the two million square miles of the icebound Arctic Ocean, which for centuries had defied man's attempts to create a Northwest Passage, were at last to be opened up. A nuclear icebreaker is to be constructed, less to create a passage to the west than to find the immense pools of oil and natural gas in the area. With the news also came the information that a Canadian expedition had found the wreck of H.M.S. *Breadalbane,* one of the ships sent to search for Franklin in 1853—125 years after she had been abandoned—preserved frozen near Beechey Island. The temperature of the water in this area, according to the expedition, stays at about 29.5 degrees Fahrenheit, and more relics are expected to be discovered.

Are *Erebus* and *Terror* still there, still guarding their secrets, hidden in the ice, part of the vast frozen wastes and now, after a century and a quarter, totally lost to sight?

2. Mary Celeste (†1872)

The greatest mystery of them all

Mary Celeste is the classic sea mystery. By comparison the many that preceded and followed her have been dwarfed by the reputation and living presence of *Mary Celeste*. What other such ship has entered the general vocabulary?

Mary Celeste was a brig, or half-brig, often called a hermaphrodite brig. She was carvel-built (her planks were flush instead of overlapping) and was brigantine-rigged, which meant that she had two masts, the foremast square-rigged and the mainmast fore-and-aft-rigged. She had a billet head—a carved scroll under the bowsprit—and a square stern. She was built in 1860 as the maiden venture of a group of pioneers on Spencer's Island, Nova Scotia, and was 99.3 feet long with a gross tonnage of 198.42. She was launched in 1861, originally named *Amazon* and registered at Parrsboro. On her maiden voyage, her master, Robert McLellan, was taken ill and died within a few days of reaching home. It was almost an omen.

For the next six years she voyaged between North American ports, England, the West Indies and the Mediterranean, but in 1867 she ran ashore in a gale at Cow Bay, Cape Breton, where she had gone to load coal for New York. She then passed through various owners until 1868, when she was transferred to American registry with the new name of *Mary Celeste*. In 1872, just before her final voyage, she became the property of a consortium consisting of James H. Winchester, Sylvester Goodwin, Daniel T. Sampson and Benjamin Spooner Briggs, of Marion, Maine. They stripped her down to her copper and at a cost of $10,000 increased her length to 103 feet, giving her a total tonnage of 282.28. She thus became a double-decker, and for the sake of easier handling, her topsail was divided into upper and lower topsails.

Benjamin Spooner Briggs, who became her new master, was born at Wareham, Massachusetts, in 1835, the second of the five sons of Captain Nathan Briggs. All Nathan Briggs's sons but one followed the sea. The elder Briggs was an affectionate father who was something of a poet and a philosopher, but he was also a spartan aboard ship who made sure his sons were granted no favors, and a strict disciplinarian who did not allow grog aboard, a rule his sons followed. Aboard ship, the sons were expected to do the same work as the crew, take their trick at the wheel, stand watch and furl sails, and to be first aloft in an emergency. In addition, they had to study regularly and recite to their father what they had learned of navigation, geography and literature. Three of the four became masters at an early age.

The Briggs family was close and affectionate, though unlucky. One son, Oliver, was lost with his ship; a daughter, Maria, was drowned in a shipwreck; another son, Henry, died at sea of yellow fever; and Nathan Briggs himself was killed by lightning at the door of his home. And Benjamin was to become the center of the sea's most celebrated mystery.

He was a quiet, retiring man, a regular reader of the Bible and a Freemason, an order he had joined while in Gibraltar. In 1862 at the age of 27 he had married a 20-year-old cousin who had been a childhood sweetheart, Sara Elizabeth Cobb, a pastor's daughter. When he took command of *Mary Celeste* in the autumn of 1872, he was 37 and had previously commanded the schooner *Forest King* (in which he and his new wife spent their honeymoon on a voyage to the Mediterranean), the bark *Arthur* and the brig *Sea Foam*. His reputation as a master mariner was well established, and, of sturdy, God-fearing New England stock, he had the highest of characters for seamanship and correctness.

2. MARY CELESTE (1872)

Contrary to later accounts, *Mary Celeste*'s crew seem to have come from much the same mold as their captain, and, contrary to the stories which suggested they numbered an unlucky 13, in fact there were only ten people aboard, including Briggs.

The first mate was Albert G. Richardson, of Stockton Springs, Maine, who was aged 28 and had served throughout the Civil War as a private in the Northern Army. He had an excellent character and had sailed with Briggs before. Briggs considered himself fortunate in being able to get him. The second mate, Andrew Gilling, who came from New York, was of Danish descent and was aged 25. The steward and cook, Edward William Head, from Williamsburg, was 23 and had just married. The other four crew members were all German from the islands of Föhr and Amaru in East Prussia. They were Volkert and Boz Lorenzen, Arian Martens, or Harbens, and Gottlieb Goodschaad, or Goodschall. According to their village records, they were all peaceable men and first-class sailors and they were all young. Briggs was sufficiently impressed to write to his mother just before he sailed that they had a good mate and steward, that the vessel was in "beautiful trim" and that he hoped they would have a good passage.

The ship was manned well within the recommendations of the Seamen's Congress held some years later in England, when a crew of seven was recommended for sailing vessels of 200 tons and a crew of nine for vessels of 300 tons; and, as he was taking his wife and two-year-old daughter, Sophia Matilda, with him, it is more than likely that Briggs selected his men with particular care. Another child, Arthur, aged seven, was left behind to start his schooling under the care of Captain Winchester, one of the part-owners of the ship.

Captain Briggs left his home about October 19 and his wife and small daughter joined him in New York on the 27th. The vessel finished loading the following weekend, November 2. Briggs was very happy with his ship but, not having been in her before, had no idea how she would sail.

On Monday, November 4, he signed the articles of agreement and the crew list at the United States Shipping Commissioner's office. On the same day the Atlantic Mutual Insurance Company, one of the five companies involved, organized insurance on *Mary Celeste*'s freight interests for $3400. Four other companies carried the insurance on the ship's hull for a total of $14,000.

The ship left her berth at Pier 44 in the East River on Tuesday morning, November 5, but anchored about a mile or so from the city because of unfavorable weather. As they waited, Mrs. Briggs wrote to her mother-in-law saying that her husband thought the crew "a pretty peaceable set this time all round." Mrs. Briggs had taken her sewing machine and her harmonium (a small foot-pedal organ) with her, together with books of music. She and her husband were both looking forward to the trip and their high spirits showed in singing sessions at the harmonium. The voyage promised to be highly profitable because *Mary Celeste* had already been chartered to carry a return cargo from Italy to New York.

With the wind light and favorable, the ship sailed on November 7, 1872, with a registered cargo of 1700 barrels of crude alcohol belonging to Meissner, Ackerman and Co., of New York, and destined for H. Mascarenhas and Co., of Genoa, for fortifying wines. The barrels were stowed in three or four tiers and were valued at $37,000.

This much is fact. But even this is misrepresented as the speculators get to work on their theories as the years pass by and *Mary Celeste* becomes *the* unique sea mystery.

The first stories said that when *Mary Celeste* was next seen in mid-Atlantic she was deserted but undamaged, with all sails set and not a rope or spar out of place. There were all the signs of peaceful occupation, work in Mrs. Briggs's sewing machine, a cat asleep, a hot meal on the cabin table with three cups of still-warm tea, a plate of porridge and a boiled egg with the top cut off. Another story stated there was a bottle of cough medicine near Mrs. Briggs's place with the cork out, yet unspilled, and more food in the galley on a still-warm stove. On the table was a child's pinafore, some religious music and books. According to another account the captain's watch was still ticking on a nail and on his table were the remains of a half-consumed meal as fresh as when it came from the galley. Another claimed that the galley range, though raked out, was still warm, seamen's clothing was hanging up to dry and the boats were undisturbed on the davits. Yet the whole ship's company had vanished. A century later, this is the story which is generally accepted.

The truth was somewhat different.

On December 14, one month and one week after *Mary Celeste* sailed, a cablegram from Gibraltar was received by the owners' agents indicating that something very strange had happened: "FOUND FOURTH AND BROUGHT HERE MARY CELESTE ABANDONED SEAWORTHY ADMIRALTY IMPOST NOTIFY ALL PARTIES TELEGRAPH OFFER OF SALVAGE."

On the day this cablegram was sent, another had been dispatched by the American Consul at Gilbraltar, Horatio J. Sprague, to the Board of Underwriters in New York: "BRIG MARY CELESTE HERE DERELICT IMPORTANT SEND POWER ATTORNEY TO CLAIM HER FROM ADMIRALTY COURT." Sprague had also informed Mr. O. M. Spencer, the American Consul in Genoa, where *Mary Celeste* was bound.

The first cable had been sent by Captain David Reed Morehouse, master of the bark *Dei Gratia* of Nova Scotia, heading from New York to Gibraltar for orders with a cargo of petroleum. The few words summed up a mystery which was to baffle everybody for over a century.

On December 4 (December 5 according to sea time) *Dei Gratia* was at a position latitude 38 degrees 20 minutes N, longitude 17 degrees 15 minutes W and 600 miles from land, her bows pointing southeast by east. She carried eight men altogether. Morehouse, her captain, was on deck at the beginning of the afternoon when the lookout reported a strange sail about four to six miles away, on their port bow. The ship was under very short canvas and the state of her sails drew their attention.

Captain Morehouse summoned the mate, Oliver Deveau, a Canadian. They studied the strange ship together through their glasses. She was making about two and a half knots, heading northwest by north toward them, and she appeared to be in distress, though no distress signal was flying. Anxious to offer assistance, Captain Morehouse hauled up and, drawing nearer, hailed her. There was no response. The strange ship was yawing and carried only a jib and a foresail set on a starboard tack, which was odd because, although the weather had been stormy and the seas were still high, it was a fine afternoon with a light northerly breeze that favored full canvas. The foresail and upper topsail had been blown away, the lower fore topsail was "hanging by the four corners" and the main staysail had been hauled down and was lying loose on the forward deckhouse across the galley chimney. All the rest of the sails were furled and, while the standing rigging seemed sound, some of the running rigging had carried away.

Deciding to investigate, Morehouse lowered a boat and Deveau

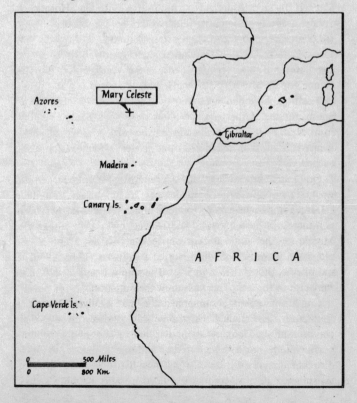

rowed across to the strange ship with the second mate, John Wright, and a seaman, John Johnson. As they passed the ship they noticed her name, *Mary Celeste*. Leaving Johnson in the boat alongside, Deveau and Wright clambered aboard and called out, but the ship was silent and appeared to be sailing herself.

There was an air of bedraggled desolation about her and a total silence except for the rhythmic creaks of the rigging, the ghostly clacking of loose blocks and the clunk and rumble of a loose cask that moved to the motion of the ship.

Wondering if the crew were all dead below of yellow fever or cholera, neither uncommon then, Deveau and Wright made a thorough search of the ship but failed to find anyone aboard. The fore hatch and the lazaret hatch were off. The binnacle, a wooden stand for the compass, had been knocked out of its place and damaged, and the compass was destroyed. The wheel had not been lashed but was undamaged, while the forward deckhouse was full of water, and there was a lot of water between decks The ship carried no boats, though it was obvious from the position of two fenders that one had been carried on the main hatch, not the proper position. She carried davits for another boat astern, but she had clearly not had a boat there because the davits had been lashed to a spar. There was nothing beyond a lifted rail to indicate how the single boat had been launched. The captain's chronometer, sextant, navigation book, ship's register and other papers were all missing.

Descending to the cabin, Deveau found everything wet through, as if there had been a great deal of water there. The deck cabin, finished in pine with a deckhead slightly raised above the deck, had six windows, all of which had been battened up with canvas and boards, though the skylight was open and raised. There was one berth in the mate's cabin and one in the second mate's, both of which appeared to have been occupied, and there was also the captain's berth, which was wide enough for his wife and child. The captain's bed was not made and there was an impression on it of a child having lain there. The bedding and clothes that were lying about were wet, and Deveau decided that the water must have come in through the windows nearby or through the skylight.

All the captain's·clothing and furniture were in order, however, and Deveau judged that a woman and child had been on board, because he found a sewing machine on a writing desk, women's clothing and a child's toys. There was nothing to eat or drink on the table and apparently a meal had not been expected. On the table alongside was a rack containing fiddles—wooden slats for holding dishes in place when the ship was moving violently in heavy seas—and the vessel's log slate. All the knives and forks appeared to be in the pantry where there was also a supply of preserved meats. Everything seemed to indicate that the occupants had left in a hurry.

A rosewood harmonium stood against the partition that divided the cabin from the saloon, with a sheet of music in place. Some of the captain's charts were in two bags under the bed, some lying loose across it. Also under the bed was a decorative cutlass-type sword in a scabbard, and bags containing clothing. The clock had been stopped and damaged by sea water and hung, minus hands, on the saloon wall.

The mate's cabin was secure. To get inside Deveau had to break away the battening and the glass of the window with a hammer from a box of carpenter's tools. Inside, on the mate's desk, were the logbook and two charts, one of which showed the track of the vessel up to November 24. There were also some letters and the mate's notebook showing the receipts for the barrels of alcohol in the hold, and on the log slate was an entry marked 8:00 A.M., November 25, showing that they had passed the island of Santa Maria in the Azores heading east-southeast.

The forward house on the upper deck was open, and the water there was up to the foot-high coaming. The sliding wooden cover in the roof of the galley, situated in the corner of the forward house, was off and the windows were shut, but everything was tidy and washed up. No cooked food was found, but there was a barrel of flour, one third used. Washing hung on a line.

There were four berths in the forecastle, the bedding again damp and "as if it had not been used." Deveau also noticed that there were only three sea chests, but he knew that sailors sometimes shared them. The men's clothing had all been left behind—

their oilskins, boots, razors, even their pipes—as if they had left in a great hurry, because a sailor would normally take such things, especially his pipe, with him. There was no sign of a message to indicate what had happened.

In the hold were barrels marked "Alcohol," apparently in good condition and properly stowed. They had not shifted and were not injured; Deveau found no other intoxicating liquor in the ship. There was no evidence of damage by fire, or of smoke or charring, and nothing to indicate that the ship had been driven over on her beam ends. Her hull appeared to be "nearly new," and she was supplied with provisions for six months, the casks all in their proper places, something that would not have been so if the ship had somehow capsized. The drinking-water casks were on deck, but their chocks appeared to have been moved as if struck by a heavy sea.

While *Dei Gratia* had passed to the north of the main group of the Azores, *Mary Celeste* appeared to have passed to the south, and Deveau estimated her present position to be 600 miles from her last marked position on the chart. With the sails she had set, she had probably come up to the wind and fallen away a little at times, but would always have kept the canvas full, and it seemed to him she had probably changed course more than once.

Meanwhile, Wright had found a sounding line made from a heaving line weighted with a bolt, which the missing crew must have been using to sound the hold. With it, he found the hold contained three and a half feet of water, a lot for such a small vessel but nothing to cause alarm because it could be cleared with a little steady pumping. Trying the pumps, simple handles attached to iron pipes bolted to the deck, Wright found they worked efficiently.

Considering the ship in good enough shape to be sailed, and since four men could handle *Dei Gratia*, Deveau returned to Morehouse and persuaded him to let him work *Mary Celeste* with a crew of three because, if they brought her to port, they would be able to claim salvage money on her. Returning to *Mary Celeste*, Deveau took with him two men, Augustus Anderson, who decided *Mary Celeste* was well capable of going "round the world with a good crew and good sails," and Charles Lund. He also transferred *Dei Gratia*'s small boat, a barometer, compass and watch, and food the ship's cook had prepared, together with his own nautical instruments.

Having climbed back aboard *Mary Celeste*, they hoisted the boat through the gap in the rail, put her into wind and used what sail was left to put her on course. Within two hours they had pumped her dry, then Deveau found a spare trysail which he rigged as a foresail and drilled holes in the forward deckhouse to drain it of water. They made no attempt to light the stove in the main cabin and, since the beds were wet, they brought out the mattresses and blankets to dry. Dividing the watches between them, they ate casked salt beef with potatoes cooked on the galley fire. Though Deveau had only a first mate's certificate, he had sailed a brig before as master and he found that *Mary Celeste*, though not much better as a sailer than *Dei Gratia,* was slightly faster. In making his own entries on the log slate, he unintentionally rubbed out some of the entries made on it by the mate of *Mary Celeste*.

At first the weather was good, so that *Dei Gratia* was able to keep them in sight as they got the ship in trim. Already Deveau was convinced that *Mary Celeste*'s crew had abandoned the ship in the belief that she had more water in her than she actually had. He was too busy to log his positions, and he knew very well that, with the men shorthanded and already tired, they might well be in danger if they had to face really bad weather. Reaching the Strait of Gibraltar, they ran into a storm, so he lay under Ceuta and later in the lee of the eastern coast of Spain. Though he had lost contact with *Dei Gratia,* he made port only a few hours behind her on the morning of November 13, when Morehouse sent his telegram to the owners.

To the weary men's surprise there were no congratulations from the authorities ashore for what was, after all, a fine feat of seamanship. Instead, as the usual notice of arrest was nailed to *Mary Celeste*'s mainmast by Mr. T. J. Vecchio, the Marshal of the Vice-Admiralty Court, Morehouse, Deveau and the crew of

Dei Gratia realized they were being regarded with suspicion.

They had become a matter of grave concern to a pernickety, suspicious and elderly bureaucrat, one Frederick Solly Flood, who was to become instrumental in confounding the story of *Mary Celeste* until it was impossible to separate truth from fiction.

Morehouse, who wished to proceed to Genoa, felt that around half the salvage money on the hull and freight interests of $17,400 and half the value of the cargo, insured in sterling at around £6522.3.0., would be his. Unfortunately Flood, who held the resounding title of Her Majesty's Advocate General and Proctor for the Queen in Her Office of Admiralty and Attorney General for Gibraltar, was in charge of the case. Born in London, he was aged 71. Full of his own importance, he destroyed at once all hopes that *Mary Celeste* would be quickly released.

In a welter of rumor, the court of inquiry sat after a lapse of five days. It was conducted by Sir James Cochrane, Commissary of the Vice-Admiralty Court, and the court persisted in behaving throughout as though *Mary Celeste* were a ship of some other name. Mate Deveau gave his testimony, concluding with his view that the crew of *Mary Celeste,* finding water in the hold, had become alarmed and abandoned the vessel, though he himself had found that she made little or no water and had assumed that all the water in her had gone down her open hatches and through the cabin. He pointed out that *Dei Gratia* had experienced stormy weather for most of her voyage, and during that period her main hatch had been off for perhaps one hour, and the fore hatch not at all. He also pointed out that he had seen no sign of damage by fire or any sign of fire or smoke in any part of the ship.

The evidence of Second Mate Wright and Seamen Lund, Anderson and Johnson followed. They did not entirely agree about whether *Mary Celeste* had a boat or not, but a yawl that she had carried amidships was certainly missing, and, judging by the lifted rail, it had probably been launched at that point. It was now known that she had not carried a longboat; it had been damaged and she had sailed without it, her stern davits, as Deveau had noticed, lashed to a spar.

Despite the slight disagreement, the general view they presented was virtually the same and their statements showed an extraordinary picture—of a well-officered, well-found and well-provisioned ship abandoned in mid-ocean for no apparent reason. It was not really surprising that Flood's suspicions were aroused. Under his instructions a survey of the ship was carried out on December 23 by Ricardo Fortunato, a diver, Vecchio, the Marshal of the Court, and John Austin, Master Surveyor of Shipping, and was attended by Flood. It lasted five hours and was remarkably thorough. Fifty of the casks of alcohol were removed to the deck and were found to be in excellent order, but Austin was unwilling to believe that *Mary Celeste* had been through bad weather as Deveau had claimed because he had noticed a phial of sewing-machine oil, a reel of cotton and a thimble balanced on a shelf. He had also noticed that the mate's bed was dry and that there were loose pieces of iron and two panes of glass in a drawer under the captain's bed, and he felt the glass would have been broken in bad weather, and considered that the shutters over the windows of the forward deckhouse would also have been damaged. But Deveau had dried the beds and he might well have replaced the articles on the shelf after they had fallen off; while the panes of glass, lying flat, might well not have been broken. Austin also reported that the galley stove was "washed out of place," and he noted a hole in the deck of the galley near the hearth which he assumed was to clear the galley of water. It had probably been made by some irritated cook paddling in sea water and was probably why the galley was not swilling with it like the rest of the deckhouse. Like Deveau, he also discounted the possibility of fire because there was not the slightest evidence of it.

While the inspection was taking place, Morehouse, eager to deliver his own cargo to Genoa, sent off Deveau with *Dei Gratia,* remaining behind himself to fight the salvage claim. At once Flood accused him of causing inconvenience. There seemed to be some official conspiracy over the abandoned ship, in fact, and not satisfied with the first examination, Flood ordered another on January 7, 1873, which he also attended. With him this time were four naval captains, Captain Fitzroy, of H.M.S. *Minotaur,* Captain Adeane, of *Agincourt,* Captain Dowell, CB, of *Hercules,* and Captain Vansittart, of *Sultan,* together with Colonel Gaffam, of the Royal Engineers. The log slate had already been seized, but Flood pointed out to the party the sheet of music on the harmonium, two ladies' hats, a nightshirt, a doll, a fan, two ladies' shawl pins, a crinoline and the phial of sewing machine oil, thimble and reel of cotton on the shelf, observing that if the bad weather Deveau had said he had experienced had really occurred, they would have fallen off. He then took the officers to the hold and suggested that one of the barrels of alcohol had been broached. Neither Deveau nor Austin had seen fit to comment on this, and one barrel out of a whole cargo was nothing to worry about, anyway.

Finally Flood pointed out a scar in the wood of the starboard topgallant rail which he suggested had been made by an axe, and indicated what he considered to be bloodstains on the deck nearby, and marks on the hull above the waterline which he felt showed that wood had been cut away.

A week later, on January 15, Captain Winchester, of Winchester and Co., managing owner of *Mary Celeste,* arrived to protect the interests of the owners and the underwriters in New York, who had empowered him to act on their behalf. Though Winchester had no doubt about the vessel's identity, the court still referred to her as "the ship supposed to be *Mary Celeste.*" In his evidence Winchester said the cargo consisted of 1701 barrels of alcohol, one more apparently than expected, with 30 tons of stone ballast beneath. She carried no other cargo and had loaded in New York's East River at a pier close to his office, so that he had made a habit of visiting her once a day during the three days she was taking the alcohol on board. He refuted the suggestions that Briggs or Richardson, the first mate, had some enterprise of their own going on, because he knew they had no private funds, though he could not be so sure that they were not carrying on some venture for somebody else.

The American Consul, Sprague, was watching the proceedings on behalf of his countrymen. In a letter to the Department of

State, Sprague said that he personally knew the missing master, who, he pointed out, had "the highest character for seamanship and honesty," and he found it difficult to account for the abandonment. Flood was by no means put off. The two surveys had strengthened rather than diminished his belief in foul play, and on January 23 he wrote to the Board of Trade in London, expressing his belief that it was impossible for *Mary Celeste* to have sailed unmanned as far as she apparently had, and suggesting that she was abandoned much farther to the east than the spot where she was found. The implication was clearly that Morehouse was a liar.

A week later the court sat again and Cochrane, announcing that a further investigation was felt to be desirable, commented that the disappearance of Mate Deveau to Genoa would be likely to influence the decision of the court. It was strange, he observed, that Deveau, who had been aboard *Mary Celeste,* should have left, while Morehouse, who had not, had remained behind.

To make matters worse, Captain Morehouse, who was represented by Mr. H. P. Pisani, was not a good witness. He was a Nova Scotian, born at Sandy Cove in 1838, who had gone to sea at the age of 16. He had become a master at 21 and had a reputation as a navigator and as a good citizen. In court, however, he was laconic and hostile, and, though this proved nothing, the suspicious Flood assumed at once that he had something to hide. Perhaps Morehouse's conduct was understandable. He was wasting time and time meant money to him, and he wasn't in the habit of taking orders. At this time also, only a few years after the American Civil War, when the United States was beginning to emerge as one of the wealthiest and most powerful nations in the world, there was a marked dislike among Britons for the American get-up-and-go spirit that he probably resented.

Flood was still convinced that he was faced with murder and mutiny, his theory—supported by the British officers who had accompanied him on board—based chiefly on the broached alcohol cask. Winchester tried to squash his suspicions by saying that captains Briggs had served under had spoken of him with warmth, but Flood was still not convinced. He clearly felt that by sending Deveau to Genoa without permission, Morehouse was deliberately removing a witness and thereby concealing information from the court. In this view he was supported by John Austin, the port surveyor, whose opinions he enclosed in a letter to the Board of Trade in London six weeks after *Mary Celeste* had been found. Flood's theory was that the crew had "got to the alcohol and in the fury of drunkenness had murdered the master . . . his wife and child and the chief mate." They had then, he suggested, damaged the bows with a view to making it look as if the ship had struck on rocks or suffered a collision, so as to induce the master of any vessel which might have picked them up, seeing the ship at a distance, to think her not worth saving. They had finally, he said, some time between November 25, the date of the last log entry, and December 5, escaped on board some vessel "bound for a North or South American port or the West Indies." Flood also drew attention to the sword found under Briggs's bed by Deveau and later by Vecchio. "It appeared to me," he said, "to exhibit traces of blood and to have been wiped before being returned to the scabbard."

Flood had no proof whatsoever for his beliefs, but they were solemnly passed on by the British government to the American administration via the British Ambassador in Washington; and the U.S. Secretary of the Treasury, William A. Richardson, circularized customs officials throughout the United States to look out for a ship carrying the alleged murderers.

By this time Sprague, the American Consul, was growing irritated with Flood's behavior, and when U.S.S. *Plymouth* arrived at Gibraltar the following week, he got Captain R. W. Shufeldt to make an inspection of *Mary Celeste*. Shufeldt made no secret of what he thought of Flood's suspicions. He rejected at once the idea of mutiny and suggested that the marks on the bows were "no more than splinters made in the bending of the planks which were afterwards forced off by the action of the sea." When the question of the broached cask of alcohol was brought up he retorted quite rightly that the crew couldn't possibly have drunk crude alcohol, which would have blinded them and might even have killed them, but either way would have rendered them incapable of mutiny or murder.

Deveau was required to give evidence again on March 4, when he returned from Genoa, and it was then that the question of the sword was raised in court for the first time. As it happened, Deveau could not produce Lund, one of the men who had been on the derelict with him. He said Lund had injured his back in Genoa and wasn't fit to appear. The mood of the court was distinctly hostile to the salvors by this time. The scar on the rail, the stained blade of the sword Deveau had found and the broached cask of alcohol had convinced the British that there had been a conspiracy on board. The Americans were equally sure that there had not: certainly all these things could have had perfectly innocent explanations. Much was made both then and later of the sword, but it was suggested that Briggs had picked it up on some foreign battlefield and he might well have intended to hang it on a wall in his home. Certainly no other weapons were found on board.

Though Deveau, curiously, was not asked about the dried beds, the articles which should have fallen off the shelf and the panes of glass in the drawer, a lot was made of a trailing halyard which he said he had found "broke and gone." He had seen no remains of a boat's painter or rope fastened to the rail of the ship, and he had not noticed the cut in the rail and had no idea how it came to be there. It appeared, he agreed, to have been done by an axe, but he felt it could not have been done while he was aboard. He also saw no bloodstains on the deck, but he and his crew had not washed or scraped the decks because there had been no time and the only water that had reached them had been washed aboard by the sea. At once Flood pointed out that the sea contained chloric acid which would dissolve particles of blood.

Asked about the sword, Deveau said he had not noticed anything remarkable about it except that it appeared to be rusty. Once again Flood interrupted. The sword seemed to obsess him, and he pointed out that it had been cleaned with lemon which had covered it with citrate of iron. This had destroyed any bloodstains it might have borne, and, anyway, the marks on it were not blood at all but some other substance put there to disguise the blood. This complicated and solemn utterance by Flood, with its grave implications, caused a sensation in the court. But it did not perturb Deveau in the slightest. He remarked that it had never occurred to him that there had been any act of violence because there was nothing to suggest there had.

Though his theory of mutiny and murder was beginning to seem unlikely, Flood refused to be beaten and started to suspect collusion. It was being suggested now that Briggs and another master, probably Morehouse, had staged a fake abandonment with the object of collecting salvage. Perhaps it was this suggestion which was picked up by American newspapers, uninhibited by such trivialities as libel laws. "ABANDONED SHIP," the *New York Sun*'s headline ran on March 12, 1873, "NO MUTINY YET BUT A SCHEME TO DEFRAUD THE INSURANCE COMPANY." Despite the indignant denials of Winchester and Co., it was openly stated that the ship was improperly cleared, had sailed under false colors and that she was to be seized on her next arrival in port. Replying in the *New York Herald* three days later, Winchester and Co. referred readers to the records of the Customs House and to the respectable insurance companies—one of which, Atlantic Mutual, still exists—which had covered the ship. *Mary Celeste* was by no means overinsured but was covered for a reasonable commercial figure and was worth more to the owners in being. They also said that the man who had started up the false hares, a deputy surveyor by the name of Abeel, was in fact apparently endeavoring to use what he claimed to be a fraudulent register to extract blackmail.

The damage to the ship's bows and the rail nevertheless remained a puzzle. Austin, the surveyor in Gibraltar, preferred to disagree with the American captain, Shufeldt, insisting that the marks on the hull could not have been made by the weather but by a sharp cutting instrument, leaving a scar about three eighths of an inch deep and about one and a quarter inches wide for a length of about six or seven feet. Both bows bore identical marks. Winchester agreed with Shufeldt and said that, though the marks had not been there when the ship had left New York, he felt the splinters had been caused when the planks had been steamed and bent during building. He could not explain the mark on the rail, which he felt had been caused by a severe blow from a sharp cutting instrument.

In February 1873 Sprague reported to the U.S. Department of State that still nothing had been heard of the missing master and crew, but that *Mary Celeste* had at last been turned over to her owners. The court's report stated only that the ship had been picked up on December 5, abandoned but in perfect condition, and brought into port. The fate of the crew was unknown and until as late as March 13, 1873, it was still expected that they would turn up or that some ship would report having found them.

By this time Captain George Blatchford, of New York, had arrived and was waiting to put a new crew aboard *Mary Celeste* and refit the ship. But Flood was still obsessed with his theories, and, though it was costing her owners money, he insisted on keeping the ship idle in port. Probably having heard the comment— even from Americans—that across the Atlantic it was always the policy to look out for a fast dollar, he was still searching for a conspiracy.

The furious Winchester had long since gone home. He wrote to Sprague from New York on March 10, saying he had left because his business was suffering while he waited in Gibraltar and because he had heard that the judge and Flood, having used up every other pretense to cause delay, were going to arrest him for hiring the crew to make away with the officers. He had deemed it wiser to leave.

Whether his belief was true or not, the suggestion of collusion was ridiculous. Winchester and Co. were a well-respected firm and Winchester himself had had a warm friendship for Briggs's family for years and was even watching over his son. Benjamin Briggs's father had also commanded a Winchester ship, and Mate Richardson had married a niece of Winchester's wife. Any suggestion that he wished them done away with just didn't make sense.

Mary Celeste was finally cleared for Genoa after a detention of 87 days and arrived there to discharge on March 21. Consul

Sprague noted in a letter to the Department of State that she had made the journey in 11 days as against the *Dei Gratia*'s 24 two months earlier. With reluctance the court awarded salvage on March 15 to the master and crew of *Dei Gratia,* but allowed only one fifth of the value of the vessel and her cargo, estimated at a total of $42,673 or a mere £1700, the costs of the suit to be paid out of the property salved. Expressing his irritation about the disappearance of Deveau, the judge also insisted that the cost of analyzing the stains on the sword and on the ship's deck should be charged against the salvors. Morehouse and his crew had expected at least a third of the value, probably even a half.

The matter finally seemed ended, but though the authorities had finished with *Mary Celeste,* the rest of the world had not, and the theories started at once. Winchester had decided that the barrels filled with alcohol in the hold had leaked gas which had exploded and blown off one of the hatch covers, and, fearing a worse explosion, the crew had escaped in the missing boat. Richardson, the Secretary of the U.S. Treasury, thought differently, however, and wrote an open letter to the *New York Times* on March 25, 1873, which clearly reflected Flood's views. He stated that the circumstances of the case aroused grave suspicion that the master, his wife and child, and perhaps the chief mate, were murdered "in a fury of drunkenness by the crew, who had evidently obtained access to the alcohol . . ." and had then, after abandoning the ship, either perished at sea or, more likely, escaped on some vessel bound for some North or South American port or the West Indies.

Mutiny was by no means farfetched. In sailing ships which took weeks and even months between ports, dislikes could grow to hatreds and there were plenty of cases of mutiny. There had been H.M.S. *Bounty* (1789), the most famous of them all, the schooner *Plattsburg* (1816), the American whaler *Globe* (1824), the brig *Vineyard* (1830) and the topsail schooner *Walter M. Towgood* (1857). And still others, after *Mary Celeste: Lennie* (1875), *Caswell* (1876), *Frank N. Thayer* (1886), *Natal* and *Ethel* (1889), *Veronica* (1902) and *White*

Rose (as late as 1908).

Crews were often a mixture of nationalities and could include Britons, Americans, Negroes, Filipinos, Chinese, Malays, French, Germans and Dutch. Intrigues and resentments were inevitable. Only the advent of radio seems to have stopped the crop of mutinies. Unfortunately, like others before and after, Richardson overlooked the fact that the crude alcohol would not have inflamed the crew, merely incapacitated or killed them.

Captain Winchester's fears about Flood's arresting him had also come to light, and barratry, a conspiracy to defraud the owners by the crew, was also offered as an explanation. The crew, it was suggested, had arranged to murder Briggs and his family and sink the ship for the insurance, but they had somehow mishandled it and lost their lives near the Azores. A waterspout was also suggested, or some submarine disturbance that had caused the abandonment, and finally that the ship had become stranded on a "ghost island," which rose and fell, and that when the crew abandoned her, the sinking of the island freed the ship.

Deserted or lost ships were far from uncommon. The French ship *Rosalie* or *Rossini* was found abandoned with all sails set in August 1840, as were the German bark *Freya* (1902) and *Carroll A. Deering* (1921). The bark *Lotta* had vanished in 1866, as had the Spanish ship *Viego* in 1868. The British cadet ship *Atalanta* disappeared in 1880 and the Italian schooner *Maramon* in 1884, even the circumnavigator Joshua Slocum with his yawl *Spray* in 1909, though he was later believed to have been run down by a ship.

A few abandoned ships drifted enormous distances: the deserted *William L. White* sailed 5000 miles in 1888; *Fannie E. Wolston* drifted halfway across the Atlantic and back again in 1891, finally foundering close to where she had first been abandoned; *Dalgonar,* in 1914, drifted 5000 miles across the Pacific. But in each of these three cases survivors appeared and there was no mystery. Those on board *Mary Celeste* were never seen again.

Other solutions followed thick and fast. *Mary Celeste* had been assaulted by a sea serpent or giant cuttlefish. The crew had fought the creature with an axe, hence the mark on the rail and the "bloodstains." That story also wasn't all that silly. There had always been reports of sea monsters. Usually they were disbelieved, but early in the nineteenth century there had been an epidemic of them. The details always seemed to be the same. In 1848 one was seen by the crew of H.M.S. *Daedalus,* but, though a Captain Smith, of *Peking,* the same year, had discounted the story by proving that what he saw on close examination was nothing more than a monstrous alga, in 1857 another creature had been seen from the deck of *Castilian,* whose captain insisted indignantly that he knew exactly what seaweed looked like and was as likely to mistake an eel for a whale as algae for serpents.

The steam dispatch boat *Electo* arrived at Tenerife in December 1862, claiming to have encountered between Madeira and Tenerife a monstrous polyp measuring 16 to 18 feet in length without counting the eight formidable arms. Attempts were made to capture it as it came alongside. It was fired at and vomited a quantity of froth, blood and glutinous matter which gave off a strong odor of musk. And the year after *Mary Celeste*'s abandonment, 1873, two fishermen off the coast of Newfoundland encountered a gigantic squid which grappled with their boat. They succeeded in severing and bringing ashore two of the tentacles, one of which was 19 feet long, though a further six feet had been destroyed. Since a further 10 feet had remained attached to the body, the monster was worked out at 44 feet. Three weeks later another monster was brought ashore near St. John's, entangled in a herring net.

But no cuttlefish could have eaten both the crew and the boat of *Mary Celeste,* and the bloodstains which had so worried Flood turned out in the end to be wine.

Pirates were next considered. Landsmen's views on pirates never change. Even when the author went to sea in the 1930s he was advised in all seriousness to beware of them. Certainly in 1832 pirates from the topsail schooner *Pinda* had captured the Salem brig *Mexican* and later an English brig, but this was the last authenticated case of piracy. By 1872 Riff pirates from North Africa, who were known to have preyed on shipping, had not been seen for years, and certainly not so far out in the Atlantic. And *Dei Gratia* had seen no sign of any other ship.

Had there been an epidemic of cholera or some other disease? This again was not impossible. In 1913 a ship was sighted off Tierra del Fuego which appeared to be totally green from masthead to waterline. This green, it was realized, was seaweed and mould. When boarded, the ship seemed to be the steamship *Marlborough*, of Glasgow, which had sailed from New Zealand in 1890 with sheep and passengers. She was literally filled with bones—those of her cargo of sheep and of her passengers. Since her log was indecipherable, it could only be assumed that an epidemic or poisoned food had killed the passengers and crew, and the sheep had starved to death.

Yet none of these theories fit the facts of the *Mary Celeste* case, and even the records do not seem to agree. Records in the American State Department refer to the American Consul at Gilbraltar as being called Johnson, not Sprague; suggest *Mary Celeste* was sailing for Naples, not Genoa; that her master was called John Hutchinson, not Briggs; and that her mate was called Henry Bilson, not Richardson.

The speculation gradually died away until a young doctor named Arthur Conan Doyle brought the whole thing suddenly to life again in 1883 with a story in *Cornhill Magazine,* called "J. Habakuk Jephson's Statement." It purported to be a true confession of a survivor of *Mary Celeste,* which Conan Doyle, adopting an error in Lloyds' List, called *Marie Celeste,* thus fixing the incorrect name firmly in legend for all time.

The story was believed. It concerned a man called Jephson, who was a graduate of Harvard University and had sailed on the ship as a passenger. He had carried a talisman given to him by an old black woman which had saved his life when the rest of the ship's company had been destroyed by a quadroon called Goring, who hated the white race.

The fussy Mr. Flood, still at Gibraltar, considered this scrap of fiction important enough to issue a statement claiming that it was pure fabrication and pointing out that people like Jephson were a menace to international relations. He clearly believed it had the elements of truth in it, however, and said he was in contact with officials in Germany, believing that some of *Mary Celeste*'s crew were hiding there after joining the mutiny which had left her derelict. Even the hardheaded Sprague, also still at Gibraltar, drew his government's attention to the story. He included a list of *Mary Celeste*'s real crew, very different from that in the story, which he considered "replete with romance of a very unlikely or exaggerated nature." Not unnaturally, the Assistant Secretary of State considered it unnecessary to pursue the matter. The story, of course, was pure fiction and had blossomed largely from Solly Flood's suspicions. Conan Doyle benefited considerably from the publicity, but the story was often accepted as fact and started a whole spate of speculation which went on for years, long after *Mary Celeste* herself was no more.

After she was released, there was difficulty in raising a crew to sail her, even to Genoa. Superstitious sailors regarded her as a hoodoo ship, and between 1873 and 1885 she changed hands seventeen times, one prospective owner, it was said, backing away at full speed when he learned her identity.

On and off for 12 years she rotted against the wharves until she was bought by a consortium from Boston, who sailed her in 1884 with a mixed cargo of food, furniture and other things. In calm, clear weather she ran aground on Roskell's Reef, near Miragoane, Haiti, on January 3, 1885. Nobody was hurt and when it was discovered she was overinsured, suspicious underwriters wanted to know more. A surveyor, Kingman Putnam, arrived to inspect the ship but discovered her entire cargo, insured for $30,000, had been salvaged and sold to U.S. Consul Mitchel in Haiti for a mere $500. What was listed as ale proved to be water, and a case listed as cutlery and insured for $1000 turned out to be dog collars worth $50. Boots and shoes were shoddy rubbers. The master and the shippers were indicted on a charge of barratry, and the Consul was also involved but escaped. At the trial the jury disagreed, but the master, Captain Gilman C. Parker, died before the case could come up again. Three months later the first mate died, then one of the guilty shippers committed suicide, and all the firms involved in the fraud went out of business. Even the ship which took the surveyor, Putnam, to Haiti and the ship which should have picked up the guilty Consul, Mitchel, came to grief.

Though *Mary Celeste* was no more, the tales and explanations about her were only just starting, fittingly enough with a theory of Solly Flood's.

In 1873 Herr A. T. Nickelsen, Chief of the Parish of Utersum, on the Isle of Föhr, had written suggesting that the effects of the two Lorenzen brothers be sent to their mother, and in 1885 the Imperial German Consul at Utersum wrote to Sprague for details about the missing men. Though the letters suggested they were perfectly normal decent people, Solly Flood, by this time over 80 years old and retired, assumed at once that the letters were written on behalf of the brothers, who, he felt, had escaped home and were wanting to know if it were safe to come out of hiding. He made this clear in a letter to Sprague which also contained a wholly irregular demand for his legal costs on behalf of Captain Briggs. Sprague's dry comment to the Assistant Secretary of State in Washington was that Flood had "always been considered as an individual of very vivid imagination, and to have survived, to some extent at least, the judicious application of his mental capacities." This, he said, was the general opinion, even among Flood's most intimate and personal friends.

In 1886 Sprague, still at Gibraltar, finally demolished Flood's theories about blood on the decks and a bloodstained sword. Asked by the Department of State for further particulars on them, he called on the Registrar of the Vice-Admiralty Court and asked to see the chemist's report. Flood had preserved this report, its seal unbroken, for 14 years, even to the extent of refusing to send a copy to the Governor of the Fortress of Gibraltar, who had made a request for it. The Registrar this time raised no arguments and the report was opened. It was quite definite in asserting that there were no bloodstains anywhere on *Mary Celeste,* and it can only be imagined that Flood, having supported so earnestly the theory of foul play, suppressed the report to save himself from embarrassment.

Once again the matter of *Mary Celeste* seemed closed, but the following year at Curaçao, in Dutch Guiana, a man of 60 claimed to have been mate of *Mary Celeste,* and in 1904, the British author J. L. Hornibrook wrote in *Chambers Journal* that the crew had been plucked from the ship, one by one, by "a huge octopus or devil fish," and recalled evidence at the inquiry of the axe-like slash on the deck rail which he felt had been caused while fighting it off. In 1911 a sailor named Mick Whelligan called on the New York police to inform them that he knew the answer. Not unnaturally, he was shown the door.

In 1913 a report in the *Nautical Magazine* by a man called Captain Lukhmanoff, a Russian, who claimed to have obtained it from a Greek, suggested that the entire crew of *Mary Celeste,* including Mrs. Briggs and the child, had been impressed to replace fever casualties on a pirate ship. The same year a group of literary men, Barry Pain, Morley Roberts, who had sailed before the mast, Horace Annesley Vachell and Arthur Robinson, got together to solve the puzzle. They were all novelists and they rejected the many supernatural theories that had been put forward as being too simple.

Pain assumed that the ship's company, including Briggs's wife and child, had been removed by a ship engaged in something shady such as piracy. Roberts felt it was a put-up job for financial gain. Vachell thought that some "unforeseen phenomenon" such as a submarine explosion had caused a lethal gas cloud into which the ship had sailed. This had driven everybody mad and caused them to jump overboard to their deaths, only the captain or the mate retaining his sanity long enough to grab the ship's chronometer and papers. Robinson suggested that one member of the crew disposed of the others by poisoned coffee and finally leaped overboard himself.

Their considerations were solemnly published in the *Strand Magazine* which a few months later also published what appeared to be a genuine account of the mystery. A highly respectable headmaster, A. Howard Linford, of Peterborough Lodge Preparatory School, in Hampstead, claimed to have had an old servant, Abel Fosdyck, who on the point of death had given into Linford's charge three boxes of papers, saying they concerned

2. MARY CELESTE (1872)

Mary Celeste. At the time the name had meant nothing to Linford, but when opened the boxes proved to contain what purported to be Fosdyck's diary for the last 30 years. With them was the photograph of a child bearing a caption in faded pencil, "Baby at the age of 2 years," said to be a portrait of Briggs's daughter, given to Fosdyck by Mrs. Briggs. According to the diary, the child had been in the habit of climbing onto the bowsprit, so a barricade, which became known as "Baby's Quarterdeck," where she could sit without danger, was rigged up. Then Briggs, who was on the verge of a nervous breakdown, challenged two of the crew to a swimming race around the ship fully dressed. To watch the contest the rest of the ship's company, Mrs. Briggs and the child crowded onto the barricade, which gave way and threw them all into the sea. Fosdyck, miraculously, was saved from drowning. The whole story was obviously fiction, but in a footnote the magazine suggested that the puzzling marks on the ship's bows might well have been made by the stays supporting the barricade. With the diary, the magazine published a facsimile of "Abel Fosdyck's" writing, but this had none of the habits of a Victorian penman such as the long "s," which could look very much like an "f," and it was not only too literate for a forecastle hand but also used terms about a ship which no sailor would ever have used. What was more, the names of the crew did not correspond to the official list, which certainly did not include the name Abel Fosdyck.

A year later, in January 1914, an article appeared in the *Liverpool Weekly Post* describing how a Mr. R. C. Greenhough, an officer in the merchant navy, had been an apprentice in the bark *Andorinha* bound for Chile in 1905. He and other men had been sent to the St. Paul Islands to collect sand and, while there, had found a skeleton and by its side a bottle containing a message. This suggested that the dead man had captained a ship engaged in some shady venture which had lost three of its crew and, coming across *Marie Celeste* becalmed, had taken her crew, including the captain. Briggs's wife and child had been shot. It was a good story, but what happened to the crew? Why did they never reappear? And how could he spend so much time writing such a long story when on the point of death?

In 1917 another version appeared. This time it came from a man called Cuppy or Chippy Russell, who it seemed was one Jack Dossel, formerly bosun of *Mary Celeste* but now living in Shrewsbury and working as an unlicensed chemist. In 1924 Captain R. Lucy, RNR, sent a story to the *Daily Express,* claiming he had heard it from another bosun and "a man called Briggs." This story suggested that *Mary Celeste* had come upon a derelict which carried a chest of £3500 in gold and silver coins. Briggs seized this, kept £1200 and gave the rest to the crew. Since they were only 50 miles from the Gulf of Cadiz, they decided to scuttle their own ship, take the derelict's boats and sail to Cadiz with the money. The appearance of another ship prevented the scuttling, but they carried out the rest of the plan. This doesn't sound at all like Briggs, who, after all, owned a third share of *Mary Celeste*'s cargo, which was much too valuable to be sacrificed for a mere £1200.

Within two years, under the name of Lee Kaye, *Chambers Journal* published an article about an elderly man called Pemberton, living in Liverpool, who claimed to know what had happened to *Mary Celeste*. Two years later a book, *The Great Mary Celeste Hoax; A Famous Mystery Exposed,* by Laurence J. Keating—obviously the same man as Lee Kaye—took up the same claim, and in 1929 the London *Evening Standard* published Pemberton's story. He was said to be then 92 years old, and the paper even published a photograph of him wearing, it claimed, the very coat he had worn when he returned to Liverpool from the incident involving *Mary Celeste,* in which he had been cook. For the first time the story involved Mrs. Briggs, whose playing of the piano—not a harmonium—in the cabin had irritated the mate, a brute called Hullock. When, during a storm, the piano fell over and crushed his wife, the frantic Briggs accused Hullock of having murdered her. Briggs then went mad and, having set fire to the ship, jumped overboard. With Hullock in command, fights broke out. One of the crew was killed, the cook, Pemberton, took over as captain and, off the Azores, Hullock and two others deserted, one of them Jack Dossel, which left Pemberton and three men aboard *Mary Celeste* when *Dei Gratia* hove in sight. According to this account, *Mary Celeste* and *Dei Gratia* had lain alongside each other in New York and Morehouse had concocted a plan to capture *Mary Celeste* and collect her salvage money.

Once again, like so many of the previous accounts, the idea was not entirely wild, because men did go mad at sea. The brig *Mary Russell* in 1828, for example, had produced a prize horror story. The ship's captain, William Stewart, was as mad as a hatter, though up to that point with a blameless record, a good husband, a capable sailor and a kind captain. Believing his crew and passengers to be plotting mutiny, he managed with the aid of three apprentices to persuade them to come and see him one by one in his cabin, where he attacked them and tied them up. Only two, the mate and another man, managed to escape, both badly wounded. During this time the ship was twice sighted, but the other ships, seeing the deserted decks and fearing piracy, gave her a wide berth. Finally the captain's mind collapsed completely so that he murdered the bound prisoners by battering them to death with a crowbar. He was just about to tie up the terrified apprentices when *Mary Stubbs*, an American schooner, appeared. Her captain was an old friend of Stewart's and, removing the two injured men to his own ship, he put two of his own men aboard *Mary Russell* and sailed in company. Believing his life to be in danger from the new crew, Stewart jumped overboard on two occasions, only to be rescued by *Mary Stubbs*. Approaching the Irish coast, he jumped overboard again and was picked up by an Irish sloop, but when this vessel approached Cork harbor, still imagining his life in danger, he jumped overboard yet again and was rescued this time by a fishing boat, whose crew happily had the sense to hand him over to the coast guard for a lunatic asylum.

Keating's story contained enough reality to be possible, though it totally ignored the fact that Morehouse was known to be an honest and cautious man and *Mary Celeste*'s crew contained no Pemberton, no Hullock and no man by any of the other names mentioned, while the cook's real name was Edward Head.

In 1924 J. G. Lockhart, a responsible writer of factual sea

stories, agreed with the notion of a religious maniac who murdered everybody, and said he had received in 1909 a letter in code—which unfortunately was never deciphered—from one Ramon Alvarado, of Cincinnati, Ohio, which claimed to solve the mystery.

Even the *British Journal of Astrology* got into the act in 1926 and found strange connections between *Mary Celeste* and the Great Pyramid of Gizeh, the lost continent of Atlantis and the British Israel Movement, and author Adam Bushey had the crew being "dematerialized" because they had sailed at a psychically vital moment over the very spot where the lost city of Atlantis had sunk beneath the waves.

But this was far from being the last of the reconstructions. In 1927 Captain J. L. Vivian Millett, of *Cutty Sark* fame and a member of the Port of London Authority, wrote in the *Journal of Commerce* that the idea that Briggs had abandoned the ship because of the danger of an explosion from the alcohol was rubbish. The explanation, he believed, was Moorish pirates.

The same year, William Adams in *John o' London's Weekly* stirred up all the doubts again by suggesting that no one really knew who had been aboard *Mary Celeste*. In fact there is little doubt. Despite the stories of brutal mates and shanghaiing which filled a great many fiction magazines of that period, ships' crews weren't really so different and the official list, made up by Mr. E. H. Jenks, Deputy United States Shipping Commissioner, gave the names exactly.

And then Hanson W. Baldwin, the American writer, made much of the strange sword found under the bed, and Commander Rupert Gould, a well-known writer and broadcaster, felt that there could not have been an explosion or a fire and considered that the crew deserted her in a panic because she was leaking badly. The panic, he felt, grew because Briggs, having died of a heart attack or something similar, was not there to curb it. He pointed out that something of a like nature had happened in Captain Cook's *Endeavour* and that in 1919 a schooner, *Marion J. Douglas,* had arrived off the Scillies as a derelict, her crew having deserted her near the Newfoundland Banks under the impression she was sinking.

In the 1930s the alcohol cargo was brought up again in a suggestion that the crew, having drunk some of it, had quarreled and died. A radio play by L. du Garde Peach suggested the crew were removed one by one by a giant octopus, and Commander A. B. Campbell, of the BBC Brains Trust, claimed in the *Sunday Dispatch* in 1942 to have solved the mystery. Dod Orsborne, the Yorkshire fishing skipper who made headlines with a tremendous journey into the Atlantic with no navigational aids other than a school atlas, felt he had found the answer when in 1942 he ran aground off Rio de Oro, in Africa. He discovered he was on a sandbank which also contained the rusty wreck of a steamer and was informed at Port Etienne that the sandbank was a "ghost island" which was produced as a result of alluvial deposits from a subterranean river beneath the Sahara. This, he felt, was what had happened to *Mary Celeste*. But, as her last recorded position was 600 miles away in the open sea, it is hard to believe.

The writer Edmond Rocher suggested that *Mary Celeste* had been lifted clear of the sea by an underwater eruption. The amazed crew had disembarked and walked around their ship but, as they did so, a second eruption caused the island that had been created to sink, and the crew were drowned. This is obviously as much fiction as the theory of Professor M. K. Jessup, instructor in astronomy at the University of Michigan, who wrote in a book, *UFO*, in 1955, that *Mary Celeste*'s people did not go down but went up, snatched up by a flying saucer.

By this point the details of the prepared meal, the sleeping cat, the warm stove and the steaming tea were firmly established, and as late as 1956 a French writer, Robert de la Croix, fell for them again, even adding a chicken frying in a pan on the stove. And with each interpretation, the legend of *Mary Celeste* was further embellished. The captain was supposed to have written in the log, "Something very strange is happening," and there was a rumor that a human bone had been found nailed to the deck. Even the names of the crew were considered to conceal an occult meaning.

De la Croix also went along with the story about the piano

falling on Mrs. Briggs, and claimed that two rafts—found on May 14, 1873, by Spanish fishermen—covered with decomposed corpses, one of which was wrapped in an American flag, were the bodies of the crew of *Mary Celeste*. He drew the conclusion that they had perished from an epidemic of some sort. De la Croix also believed that *Mary Celeste* had been sighted by a steamer only an hour before *Dei Gratia* spoke her and that there were men aboard at the time, claiming that Morehouse, for some nefarious reason of his own, had in fact put three of his crew aboard *Mary Celeste* before she left New York.

It was de la Croix's new claim that just before the *Mary Celeste* incident, Morehouse had come across the three-masted bark *Julia,* which was a derelict, with the remains of a meal on the table, with stewed fruit, bread and a few bones, and it was this that had suggested the idea of the subsequent fraud, and that his men were aboard when he found *Mary Celeste,* having already gotten rid of the rest of the crew.

Even the Bermuda Triangle became involved in 1975, when Lawrence Kusche, trying to solve the mystery of disappearing ships in that area of the ocean bounded by Florida, Bermuda and Puerto Rico which has seen so many unexplained disasters, noticed that *Mary Celeste* was also found not so far from its boundaries. His stories sprang largely from the American side of the Atlantic. He quoted the story of a drunken mutiny first offered by Richardson, the U.S. Secretary of the Treasury, in the *New York Times,* and all the other suspicions prevalent in the States at the time such as barratry, murder, waterspouts, submarine disturbances and rising and falling ghost islands.

The British writer MacDonald Hastings, armed with a book of press clippings and the view of a Captain T. E. Elwell, who had written an article of his own on *Mary Celeste* for *Chambers Journal* in the 1920s, also had a go at solving the mystery. He stuck carefully to facts. A novel featuring the mystery appeared in 1978, and the latest author to write on the subject, John Maxwell, in a fictionalized version published in 1979, used much of Flood's court behavior and produced an imaginative—and not far from correct—account of what might have happened on board. On December 16, 1979, on BBC-TV the author Rupert Furneaux claimed to have found a clue to the mystery. It was based on the statement in Conan Doyle's story of J. Habakuk Jephson that *Mary Celeste*'s boat was still aboard when she was found. Unfortunately, this was no clue because the fact that it was not had been known from the day *Mary Celeste* arrived in Gibraltar.

Perhaps the only really safe account among this myriad of speculations and interpretations was that of Charles Edey Fay, who died in 1957. He was an underwriter and senior executive of Atlantic Mutual, one of the five companies which insured the ship. His book, published in 1942, stuck rigidly to facts.

When *Mary Celeste* reached Santa Maria, the southernmost island in the Azores, he wrote, she had endured very bad weather and her crew were suffering considerable discomfort. She slipped along the north coast of the island because Briggs, he felt, like Columbus before him, was aware that by going to the south they would fall within the limits of the trade winds from the northeast, which would put the vessel off her route to Gibraltar. Nevertheless the northerly route past the island carried them near the dangerous Dollabarat Shoal, so he wondered if Briggs could have been ill or incapacitated by an accident and someone else was running the ship. They passed the eastern end of the island some time in the forenoon of Monday, November 25—a time Fay fixed by the last slate-log record, by the absence of any preparations for a meal and by the fact that the captain's bed was unmade, something Mrs. Briggs would not have allowed for long. The crew up to this time had not taken off the hatch covers because of the weather, so that there had been little ventilation at a time when atmospheric changes had been producing some effect on the alcohol in the hold. According to the Servicio Meteorologico dos Açores, the weather had now become calm, however, and Captain Briggs ordered the opening of the hatch. It was at this moment that something happened.

From this point Fay began to use the theory of Dr. Oliver W. Cobb, of Easthampton, Massachusetts, a cousin of both Briggs and his wife and a man who had also served at sea. Cobb had drawn attention to the fact that *Daisy Boynton,* on a voyage to Spain in 1930 under the command of Captain Henry C. Appleby, with a cargo of petroleum, had had her hatches blown off.

Dr. Cobb, while helping the mate of the brigantine *Julia A. Hallock* to check her 1600 barrels of petroleum, noticed that some of the metal hoops around the barrels were bright, as if they had been subject to chafing in the hold. Despite the custom of careful loading and wedging with wooden billets, they had still clearly rubbed together so that it was possible that a similar rubbing had taken place in the hold of *Mary Celeste* and that even a spark and an explosion had occurred.

Cobb pointed out that nine of the barrels in *Mary Celeste's* hold were empty, a leakage that was nothing to worry about in a cargo of 1701 barrels, but his theory was that the uprush of fumes from the unventilated hold may have been strong enough to alarm the crew, and might well have been accompanied by a rumbling or roaring noise, as if the ship were about to blow up. The fore hatch cover was found by Deveau lying on the deck, on the port side about three feet from the hatch, but he had felt it could not have been flung there by an explosion because no sign of one had been found.

According to Dr. Cobb, Briggs was alarmed at the possibility of an explosion and gave the order to abandon ship. The lazaret, where the rope was stored, was opened, but it was then decided that instead of using a new stiff rope, which would be difficult to handle in a hurry, it would be wiser to use the main peak halyard, a stout rope about three inches around and about 300 feet long. In the ship's boat at the end of this line, the crew would have felt safe and been able to return to the ship if the danger passed. They would have gathered a few necessities such as food and drinking water and, as they drew away, they would be able to see the mainsail lying on top of the forward deckhouse. It had been in the process of being let down when the alarm had come and it had been allowed to go.

The calm of the morning, as shown in the records of the Servicio Meteorologico dos Açores, was followed by a sudden squall. As the ship lunged forward under it, the long towline parted—or

more likely a faulty, hurriedly tied knot in the heavy rope gave—
and the occupants of the boat, only 16 to 20 feet long with less
than a foot of freeboard, found themselves adrift. According to
the records, there was also a heavy rainstorm in the area which
must have added to the difficulties. The nearest land, at Santa
Maria, offered little accommodation with its precipitous coast,
and it is probable that the boat was blown southeastward and
overwhelmed by the sea.

Cobb also squashed Flood's theory that no ship could sail so
far as *Mary Celeste* appeared to have sailed. He quoted the case
of the schooner *William L. White,* abandoned off the Delaware
in 1888, which sailed for 10 months and 10 days, a distance of
more than 5000 miles, an average of 32 miles a day or at an
average speed of one and a third knots. It did not seem impossible for *Mary Celeste* to have sailed 750 miles in 10 days, but
Cobb felt she had sailed before every change of wind with three
sails drawing, the foretopmast staysail and the jib keeping her on
course by preventing her from coming into wind. He pointed out
that according to records from November 25 to December 5,
1872, northerly winds had prevailed and the speed would be
three to four miles an hour. *Mary Celeste* probably went easterly
at two and a half miles an hour for nearly eight days, or 480
miles. Then, in a change of wind, she shipped a sea, which filled
the forward deckhouse and the hold, lost her foresail and upper
topsail and then went westerly at two miles an hour for 96 miles,
which would bring her approximately to the position where she
was discovered.

Such a detailed argument made sense, but it didn't entirely
solve the mystery, and certainly Flood's belief in collusion cannot
be entirely discounted because J. G. Lockhart insisted that Morehouse's wife had told him that Morehouse and Briggs were old
acquaintances and had dined together at the Old Astor House
in New York the night before *Mary Celeste* sailed. The American
writer Hanson W. Baldwin also claimed in 1956 that Briggs and
Morehouse were old friends and often met in different ports of
the world. If this was so, with both of them occasionally taking
their wives along with them, Mrs. Morehouse would certainly

2. Mary Celeste (1872)

have known both Briggs and his wife well, and she could hardly have been mistaken. It is quite possible that the two masters did see something of each other while loading, but there seems to be nothing but hearsay to back up the claim that they were close friends and no question was asked about it during the hearing of the salvage claim.

Nor did Morehouse mention it in his evidence, or mention recognizing *Mary Celeste,* as he must have done had he known her captain. Neither did he mention it to his crew when she was sighted, and surely that would have been a natural first reaction. Perhaps the friendship, like the dinner, was just another of the *Mary Celeste* myths.

Mrs. Frances Richardson, the widow of Albert Richardson, the missing mate, always believed her husband was the victim of murder and mutiny, and said as much to reporters on March 9, 1902. Richardson's sister, Mrs. Priscilla Richardson Shelton, thought the same; and his brother, Captain Lyman Richardson, thought they had been killed by the crew of *Dei Gratia.*

Deveau, who became a captain, and died in 1912, believed to the end of his days, as he had said in evidence, that the crew abandoned *Mary Celeste* because they thought she had more water in her than she actually had. Morehouse believed the ship was becalmed just north of the dangerous coast of Santa Maria in the Azores and, sensing danger, the crew took to the boat but failed to attach a line to the ship. When a breeze sprang up, it was the boat which was driven ashore and the ship which escaped. Briggs's brother also believed this theory.

Winchester, on the other hand, took Cobb's view of the danger of exploding alcohol and the crew making the boat fast to the peak halyard. Yet it still remains a puzzle why an experienced captain should lose his nerve and make such a mess of the abandonment. Perhaps having his wife and child with him made Briggs act too hastily, because he must have known that in the middle of the Atlantic *Mary Celeste* was safer than an open boat.

An even more significant attempt was made to solve the mystery in 1967, this time by Sir William Charles Crocker. Crocker was an insurance expert like Edey Fay. He was also a member of

Lloyds and something of a detective because of his experience in dealing with shady fire insurance claims. Having an American wife and knowing Edey Fay, he investigated the mystery of *Mary Celeste* for his own amusement, starting from the simple conviction that a master as experienced as Briggs would not have abandoned his ship merely because of an uprush of air or the fear of an explosion. What happened, he felt, must have been much more frightening. And so, despite the absence of charring or scorching, he settled firmly for the theory of a threat of fire or an explosion.

Like Fay, Crocker noticed that the meteorological records at the time of the abandonment indicated a calm sea, something which the open cabin skylight seemed to confirm, but that there had just been a period of bad weather which had caused the crew to batten the ship down. They had just started to open up the ship by raising the skylight when whatever happened to cause the abandonment occurred.

Most of the reputable and experienced investigators had all along suspected an explosion in the hold and the fear of another more destructive explosion. Yet the most searching examinations had failed to reveal any sign of such an explosion. However, the loss of the highly volatile alcohol during the voyage, equivalent to nine barrels, was undoubtedly due to vaporization and was accepted by the consignees as perfectly normal. But this mixture in the hold of air and alcohol vapor, Crocker pointed out, could be highly explosive when the "alcohol vapour represented not less than 3 per cent and not more than 14 per cent by weight." Beyond those limits at either end the mixture would not fire, but within those limits the mixture would release enough power to send a ship sky-high. Briggs must have been alive to this danger and to the need to open the hatches occasionally for ventilation. If, however, there had earlier been bad weather—suggested by the battened deckhouse windows and the meteorological records in the Azores—which had prevented this, he may have grown acutely apprehensive. According to Crocker, Briggs was navigating a potential bomb!

Above the hold was the deckhouse and the galley where the

cooking was done. Connecting the galley with the hold was the aperture which the surveyor, Austin, noted "in the deck near the hearth." He and his fellow inspectors regarded it merely as a means of emptying the galley of sea water. But, said Crocker, "its resemblance to the touch-hole of a cannon escaped notice." With the hold full of alcohol, it was highly probable that the vapor creeping up through it would sooner or later be lit by a cinder or the flames of the stove and flash back to the lethal cargo below. Crocker believed this is what happened and that the resulting detonation was sufficiently alarming, with Briggs already apprehensive after the bad weather had forced him to keep the hatches closed, to convince him that worse was to come. Riding astern in the ship's boat at a safe distance either until the explosion took place or until ventilation through the hatch made it safe to return seemed the most intelligent thing to do.

The first petroleum well in the United States had been drilled in 1859, and long before 1872 ships carrying that spirit had been known to blow up. There was no mystery to them, however, because there had been survivors, and the resulting havoc and marks of burning were clear. With this in mind the experts who examined *Mary Celeste* and the would-be experts who wrote about her later decided that, since there were no signs of explosion or fire in *Mary Celeste,* there could not have been one. Every intelligent examiner of the mystery suspected that an explosion of alcohol fumes had something to do with the disaster, but everyone was puzzled by the absence of burn marks. Crocker, with his experience of many fire insurance claims, pointed out that unlike petroleum-air mixtures which, on exploding, generate carbon and leave ample signs in the form of soot, alcohol, containing oxygen, burns with a nonluminous flame. Consequently its explosion rarely leaves the smoky signs of burning the inspectors sought. It was this explosion, Crocker believed, that had even caused the marks on the bow. Shufeldt had been right in surmising that the planks had splintered during the building, but it was an explosion in the main hold, delivering its punch well forward and three feet above the waterline, that had finally forced off the splinters. Perhaps also it was this explosion that disturbed the galley stove which Austin found "out of place."

When Deveau and his two companions boarded the ship, the danger from the touchhole near the stove no longer existed and they safely lit the stove to cook their potatoes. After several days and many miles, the leaked alcohol vapor in the hold must have dispersed. If the wind was strong enough to blow the sails to rags, it was surely strong enough to clear the hold of gas, in the same way that modern fuel carriers ventilate their tanks after discharging.

Sir William Crocker's explanation of the mystery of *Mary Celeste* is the best to date.

3. U.S.S. Maine (†1898)

Disaster in Cuba

In addition to ships which have disappeared and ships which have lost their crews, there have also been ships which have started wars.

Rebecca, Captain Robert Jenkins's ship, was stopped in 1731 by Spanish coast guards determined to prevent British merchantmen from obtaining a share of their New World trade, and during the ensuing attempt to defend his cargo Jenkins lost an ear. It was purportedly this ear, produced before a wildly patriotic House of Commons in 1738, that so incensed the British it provoked a minor war between Britain and Spain in 1739, which a year later drew Britain into the War of the Austrian Succession and eventually the Seven Years' War.

During the American Civil War the British-built Confederate raider *Alabama,* carrying English guns and English sailors and in action against Northern ships, all but sparked off a rupture between the Federal government and the British.

In 1915 the liner *Lusitania* was sunk off the Old Head of Kinsale, Ireland, by a U-boat with the loss of American lives, an event which as much as anything brought the United States into the First World War in 1917. The Assistant Secretary of the United States Navy at the time was Franklin Delano Roosevelt. By 1941 he was President of his country and a great friend to Britain. He had promised Winston Churchill every help he could, but finding the American people reluctant to join a European squabble, he indulged in a little scheme to make them change their minds. He had the ancient schooner *Lanikai*, which had been used in 1937 by Dorothy Lamour and Jon Hall in the film *Hurricane*, commissioned as a warship, and ordered her out to trail her coat in the hope that the increasingly aggressive Japanese would fire on her, thus committing an act of war which would enable him to enter the conflict against the Axis powers. *Lanikai*, unlike a real warship, would be no loss.

As it happened, the Japanese struck at Pearl Harbor before *Lanikai* was spotted, but *Lanikai*'s captain, Lieutenant (later Rear Admiral) Kemp Tolley, always believed that Roosevelt was inviting war. It certainly seems likely because Roosevelt would have been well aware that a ship had brought the United States into a conflict with Spain long before *Lusitania*, and perhaps it was no coincidence that some of the guns placed aboard *Lanikai* had been used in the Spanish-American War of 1898.

It had always been one of the staunch beliefs of the United States, stated firmly in the Monroe Doctrine, that the rest of the world should keep its hands off the American hemisphere. It had been invoked during the nineteenth-century revolts in South America to throw off the yoke of European domination; it was strongly invoked against France in 1865 when France was trying to set up a Mexican empire with a Habsburg prince as emperor; and it was invoked again in 1898 against the Spanish in Cuba.

Columbus had discovered Cuba in 1492 and it had remained a colony of Spain ever since. Even as early as 1863, Americans had felt that the United States would be justified in wresting the island from Spain, and that the time would probably be ripe after the Civil War when the United States would be the most powerful military nation in the world. The struggle for Cuban independence had been going on ever since 1868, and by 1895 this movement had ended in overt rebellion.

There was great support for the rebels in the United States. The American press—especially William Randolph Hearst's *New York Journal*—was particularly vitriolic against the Spanish, and published articles by moderate Spaniards in Havana, criticizing their home government. In the late summer of 1897, however, when the Spanish premier, Antonio Cánovas del Castillo, was assassinated, the prospects of peace brightened because his successor, Praxedes Mateo Sagasta, was against war in Cuba and, largely because of American pressure, 59-year-old General Valeriano Weyler y Nicolau, the Captain-General of Cuba, was recalled and a measure of autonomy granted to Cuba. Weyler had been the hero of the Spanish military party as he was strong, tough and cruel to imprisoned rebels. The army resented his recall bitterly and considered it a humiliating concession to the United States.

Though tensions eased, there was no guarantee that Spanish efforts to placate Cuba would be successful. Yet, though the Cubans were not strong enough to win, they were also not weak enough to capitulate, and by 1898 the island was desolate, with much of its population dead, diseased or starving. And with yellow fever spreading across the Gulf of Mexico and the Straits of Florida, war fever was rising among the Americans.

The nation was restless. The growth of the technical age and the power struggle already hung a shadow over the world that would culminate in 1914. The Americans, idealistic, ambitious and secure in their isolation, were actually looking forward to a war, while the Republican Party, then in power, knew that if they permitted the independence struggle to continue without helping the Cubans, they would be defeated in the next election.

On January 12, 1898, there were riots in Havana, provoked ominously by Spanish officers who had arrived in Cuba to fight the insurrection, and they had been directed chiefly against those Havana newspapers which favored autonomy. The young Spaniards had been stung by the insults against the Spanish army, but no one had been seriously injured and no property destroyed except in one or two newspaper offices. Admittedly there had been a lot of wild talk, threats and insults, of which too much was made in the American press, particularly by Hearst, and as a result the second-class United States cruiser *Maine* was ordered from Norfolk, Virginia.

3. U.S.S. MAINE (1898)

Maine had originally been designed as an armored cruiser with squaresails and topgallantsails, but later the sails were abandoned and she was styled a second-class battleship. She was designed at the Navy Yard, New York, and launched on November 18, 1890, commissioned in 1895 and left the yard the same year. She had a turret on either side, projecting from the hull, with two 10-inch guns in each, and in addition she carried six six-inchers, seven six-pounders, eight one-pounder rapid-firers and four torpedo tubes. She was 319 feet long, with a 57-foot beam, a displacement of 6682 tons, 9290 horsepower and a speed of 17.45 knots. She had an armored belt extending for 180 feet at the waterline on each side, but, though virtually a new ship, by 1898 she was already out of date.

To modern eyes she had an unusual appearance. Painted in the peacetime colors of white hull, straw-colored superstructure, masts and stacks, with black guns, she differed greatly from other vessels in having three superstructures instead of the usual one. She had ample quarters, though they were rather hot in warm weather, the crew being berthed chiefly in the forward and central superstructures and the officers aft. The breaks in the superstructure were to allow the turrets to rotate and fire across the deck.

Her commanding officer, Captain Charles D. Sigsbee, was 53 years old. Born in New York, he had graduated from the U.S. Naval Academy and seen service in the Civil War. He prided himself on his shiphandling and spent years on hydrographic work, gaining an international reputation. He had one blemish on his record. In 1886 a board inspecting *Kearsage,* the Civil War veteran he then commanded, found that she was dirty and that Sigsbee had failed to comply with ordnance instructions. He explained the deficiencies as being due to bad weather and the age of the ship, and the criticism did not affect his career.

He was the second commanding officer of *Maine,* having taken over in 1897 a few weeks after being promoted to captain. Early in his command he took the ship into New York harbor without a pilot and, proceeding down the East River, in a dangerous situation avoided ramming a crowded excursion steamer only by ramming a pier instead. At least one high-ranking officer considered he had used poor judgment by operating in crowded and restricted waters without a pilot, but his decision to hit the pier instead of the steamer won him a letter of commendation.

Great secrecy was maintained and an elaborate code was used in the arrangements for *Maine*'s departure for Cuba. It was generally believed that the ship was due to go to New York, and the message to leave for Havana was eventually brought by torpedo boat while she was off Tortugas Roads with the North American Squadron. The excuse given was that the United States government had decided to resume the friendly naval visits to Cuban ports that had been customary before the 1895 insurrection. The truth was that the U.S. Consul General in Havana, General Fitzhugh Lee, hearing of anti-American plots, had urged protection for American nationals. Lee was a nephew of the great Robert E. Lee and had been a distinguished Confederate cavalry general during the Civil War. Blunt and outspoken, he believed Spain would never mend her ways, and yet he had no faith in the insurrectionists' ability to form a government. He was convinced that the United States would have to intervene eventually and he wanted naval vessels available.

Theodore Roosevelt, Assistant Secretary of the Navy, a man of great energy and enthusiasm, was certain that the riots meant war was near; Alvey A. Adee, Second Assistant Secretary of State, felt Cuba was in danger of being plunged into chaos; and William R. Day, Assistant Secretary of State, was in agreement. President William McKinley came to the conclusion that the situation was so explosive as to be unpredictable.

Enrique Dupuy de Lôme, the Spanish Minister in Washington, a shrewd professional diplomat, had been optimistic until the riots. Echoing the words of Maria Cristina, the Queen Regent of Spain, he felt that it was only the American people who were keeping the insurrection in Cuba alive. He considered Lee's outspoken views proclaimed the failure of autonomy, and he had warned that the presence of an American naval vessel would be regarded as an unfriendly act.

Various pretexts, such as a need for fuel or the sending of dispatches to Washington, had been suggested as a reason for *Maine*'s entering Havana, and eventually the solution of restarting the friendly visits of warships stopped by McKinley's predecessor, President Grover Cleveland, was settled on. All Dupuy would say to this was that, since America and Spain were at peace, surely the visits should never have stopped. In fact, when the news came that *Maine* was to arrive, Lee had to admit that the authorities in Cuba were not hoodwinked, and the speed with which the ship was sent hardly conformed to customary international law.

Uncertain of his date of sailing but knowing he would have to respond quickly if he received a summons from Lee, Sigsbee had taken coal aboard at Newport News between November 21 and 23. It was bituminous coal, susceptible to spontaneous combustion but with better burning qualities than the anthracite he had received at Key West, and he was careful to hoard it against eventualities. Not knowing what his reception would be, he also took care to have ammunition handy for his guns.

By mid-morning on January 25, *Maine,* her flags flying, was lying off the Cuban capital. Asking the pilot, Julian García Lopez, what the reception would be like, Sigsbee was told that if the Americans behaved themselves they had nothing to fear. Lopez also showed Sigsbee a chart of the harbor and the proposed mooring and asked if he was happy with it. Sigsbee was.

Throngs of people crowded the waterfront to watch as the ship came to her buoy. Though knowing that all nations protected their harbors with mines, nobody aboard was worried. To the harbor officials, however, the arrival of *Maine* looked increasingly like a hasty act. Concerned with the danger of yellow fever, they could not understand why Sigsbee was not prepared to offer the proper documents showing he had a clean bill of health, and they considered that without them the ship should have been placed in quarantine.

There were two other men-of-war under the guns of Morro Castle at Havana, the Spanish ship *Alfonso XII* and the square-rigged German training ship *Gneisenau. Maine* passed between them to reach her position. While the Hearst press trumpeted, "Our Flag in Havana at Last" and urged that American vessels occupy all Cuban ports, Spain reacted by arranging to send the cruiser *Vizcaya* to New York, on what was also termed a friendly visit. Under the circumstances, the Americans could do nothing but accept the suggestion with good grace.

In Havana, the Spanish considered *Maine*'s presence an irritation, and America's attitude toward the insurrection—one of sympathy without giving any active support—was deeply resented. But both the Spanish and the Americans were rigidly polite. Visits and salutes were exchanged, and bullfight tickets and a case of sherry were sent aboard *Maine*. Nevertheless, Dupuy de Lôme pointed out that Sigsbee, though punctilious in exchanging courtesies, completely neglected the autonomous government, and prodded by Washington, he was obliged to repair the omission. But when Sigsbee and his officers returned from lunch with Lee, they noticed pamphlets being distributed.

One of them reached Sigsbee. It made no bones about its intention.

Spaniards,
LONG LIVE SPAIN WITH HONOR!
What are you doing that you allow yourselves to be insulted in this way? Do you not see what they have done to us in withdrawing our brave and beloved Weyler, who by this very time would have finished with this unworthy rebellious rabble who are trampling on our flag and on our honor.

Autonomy is imposed on us, to cast us aside and give places of honor and authority to those who initiated this rebellion, these low-bred autonomists, ungrateful sons of our beloved country!

And, finally, these Yankee pigs who meddle in our affairs, humiliating us to the last degree, and, as a still greater taunt, order to us a man-of-war of their rotten squadron, after insulting us in their newspapers with articles sent from our own home!

Spaniards! the moment for action is come. Do not go to sleep! Let us show these vile traitors that we have not yet

lost our pride, and that we know how to protest with the
vigor befitting a nation worthy and strong, as our Spain is
and always will be.

Death to the Americans!
Death to Autonomy!
Long Live Spain!
Long live Weyler!

Sigsbee thought little of the circular, because such things were
not uncommon in Havana. But later, as he and his officers stood
on the quarterdeck, a passenger ferry passed with derisive whis-
tles and calls—though, Sigsbee felt sure, not from Spanish officers or soldiers. With most of Havana Spanish or Spanish-Cuban,
he considered the locals behaved with restraint.

Somehow, however, the *New York Journal* obtained a private
letter written by Dupuy de Lôme in which he had described
McKinley as "a coarse politician" and suggested that autonomy
was a sham. To the American public, it was further evidence of
Spain's duplicity, and beneath the veneer of official courtesy, the
Spanish and Americans at Havana began to eye each other war-
ily. Sigsbee was ordered not to allow the crew ashore, but Madrid ordered Havana to avoid trouble if he did and within a few
days the situation seemed calm enough to permit the officers at
least to visit the city. For the crew the novelty of being in harbor
soon wore off. They were crowded and the ship was ill-ventilated,
and even the bumboats shunned *Maine*. Lee was happy, however.
The situation seemed to be under control, and he felt there was no
danger from yellow fever before April or May and that, now that
the Spanish had gotten used to seeing an American ship, others
should follow.

Despite the easing of tension, Sigsbee took every precaution,
with sentries on the forecastle and poop and ammunition for the
rapid-firers handy. Every visitor aboard was carefully watched,
but, though the Spanish appeared friendly, it was noticed on one
occasion that, instead of the group of Spanish officers who were
expected, only one arrived. Sigsbee called on Admiral Vicente
Manterola, the naval commander-in-chief, and on General Ra-

món Blanco y Erenas, who had just succeeded Weyler as Captain-General, assuring them of American friendship. He received a visit in return.

On the night of February 15, 1898, when *Maine* had been in Havana for three weeks, there was no wind and, with the ships lying in all directions, it was noticed aboard *Maine* that she was pointing toward the Machina, the Spanish admiral's palace, a direction she had not taken up before and a position she would need to take to open fire on the shore fortifications. With a turret on either side instead of in front of the bridge, *Maine* could not bring her full fire to bear unless her bows were pointing at her target.

The night was dark, overcast, hot and sultry. Almost all the officers were aboard, together with all the crew, 328 in number, and apart from the men on watch they were all turned in and all watertight compartments (there were 214 altogether) not occupied were closed. Sigsbee was in his cabin aft and his mess attendant, James Pinckney, had brought him a thin coat in place of his heavy uniform jacket. A Marine, Bugler C. H. Newton, blew taps and Sigsbee stopped to listen. Then at 9:40 P.M. just as he was putting the finishing touches to a letter, there was a report, resembling a gunshot, followed by a tremendous explosion. Sigsbee described it as a "bursting, rending and crashing sound or roar of immense volume, largely metallic in character . . . followed by a succession of heavy, ominous, metallic sounds, probably caused by the overturning of the central superstructure and falling debris." The lights went out.

Lieutenant John Hood was sitting on the deck when he heard what he thought was an underwater explosion. There was a second explosion. and he saw the whole starboard side of the deck and everything above it "spring into the air." As the deck lifted, bent almost double, it brought down the funnels, and when the debris stopped flying he saw a mass of foaming water, filled with wreckage and groaning men, in an area around the ship of about 100 yards in diameter, while on the quarterdeck a man was pinned down by a fallen ventilator. It was Hood's view that the

3. U.S.S. MAINE (1898)

first explosion had done all the damage and the second merely added to it.

Other survivors described "a trembling and buckling of the decks, and then a prolonged roar." Men in the junior officers' mess heard a dull report, followed by the disappearance of lights, and started groping their way out through a cloud of steam and a tremendous rush of water and burning cellulose that flared across their path. Cadet Wat T. Cluverius, who later gave one of the clearest testimonies of events, noted that his first knowledge of anything odd happening was a slight shock, as if a six-pounder gun had been fired somewhere about the deck. After that, there was "a very great vibration" in his cabin which was followed by a very heavy shock, continued vibration, the rushing of water through the junior officers' mess room and the sound of something breaking up.

To Lieutenant George F. M. Holman the explosion was precisely similar to many other submarine explosions he had heard, except that it was on a much larger scale. Fireman William Gatrell, who was in the steering engine room, lower in the ship than anyone else who escaped, saw a blue flame as the explosion occurred. All the men with him were lost, but before he fainted he managed to struggle to the deck, where someone grabbed him and threw him overboard.

One point was clear. The explosion had occurred in the fore part of the ship. The farther forward the survivors were, the more severely they felt the explosion. Some had miraculous escapes. A Marine corporal who was lying in his hammock on deck was blown clean through the awning above him. Another man, standing by the after funnel, saw a puff of smoke, and was lifted into the air to come down on the quarterdeck some 40 feet away. Men were flung violently about, blown in all directions "like scraps of paper in a gale of wind," burned, choked or half-drowned, yet they somehow managed to crawl, scramble or swim to safety.

Probably only two men escaped from the berth deck forward where most of the men slept. One of them, Bosun's Mate Charles

3. U.S.S. MAINE (1898)

Bergman, was flung somewhere he couldn't identify, but it was so hot that it burned his legs and arms and his mouth filled with ashes. The next thing he knew he was in the water with wreckage dragging him down. He fought clear, only to be borne down by more wreckage, and escaped because he was carried to the bottom where the wreckage subsided on either side of him. The other man who escaped, Jeremiah Shea, could only say he owed his escape to the fact that he must have been "an armor-piercing projectile."

From on shore a Spanish officer, Lieutenant Julio Peres y Perera, saw an enormous blaze of fire rise up and *Maine* almost entirely covered by black smoke, from which many colored lights were darting away into space. Fire followed and he watched the bow sink. Others saw a mighty column of flame and smoke, and when, some minutes later, it dispersed, all that could be seen of the American ship was a dark, huddled heap of wreckage, with one end burning furiously and the other crowded with men.

As the ship listed to port and began to sink, Sigsbee began to grope his way through the blackness and smoke. Private William Anthony, a Marine, guided him to the deck where a few officers and men had collected. Because the main deck was sinking rapidly, he made his way to the poop, which by this time was almost all that remained above water. Neither he nor his officers were yet aware of the great loss of life, and since, he said later, "there was the sound of many voices from the shore, suggestive of cheers," he considered the possibility that the Spanish had declared war and might attack, and sentries were set.

Everybody was behaving well in a demoralizing situation. No one knew what had happened and, like Sigsbee, most of them thought *Maine* had been blown up by the Spanish and were ready to put up a fight. The flames were increasing and an attempt was made to bring hoses to bear on them, but the fire mains had been destroyed and there were no men available. Then, as the extent of the damage became clear, they heard cries for help and saw in the water men who had jumped overboard to avoid the flames or been hurled overboard by the explosion. Boats were lowered, but only three were available out of

3. U.S.S. MAINE (1898)

the 15 which the ship carried. *Alfonso XII* and *City of Washington,* which were lying to port and starboard, were helping and the Spanish did all that could be expected. The fire grew fiercer and spare ammunition in the pilot house or thrown out of the magazines began to explode.

Waiting until he was satisfied that *Maine* was resting on the bottom but expecting the forward 10-inch magazine to explode at any moment, Sigsbee ordered the magazines flooded. As it happened, the magazines were flooding themselves. Sigsbee then ordered the men with him to get into the boats. By this time the ship was so low in the water it was possible to step straight from the deck. Sigsbee was the last to leave. The boats moved to *City of Washington,* warning other boats arriving at *Maine* to keep clear in case of a further explosion. Wounded men were taken aboard *City of Washington, Alfonso XII* and a Spanish transport, *Colon.* The work went on most of the night and the treatment of the wounded by the Spanish was humane and kind.

Meanwhile, in other ships nearby, men recovering from the shock of the explosion were trying to assess what had happened. Captain F. G. Teasdale, of the British bark *Deva,* thought that a ship had collided with him. There seemed to be a shot, then, two or three seconds afterwards, an explosion. Rushing on deck, he saw debris from *Maine* still going up into the air for about 160 feet, before it separated and came down in small, glittering, burning pieces. Fragments continued to fall on his ship for some time.

Mr. Sigmund Rothschild, a passenger in *City of Washington,* heard what he thought was a shot, then saw the bow of *Maine* rise out of the water. A few seconds later in the center of the ship there was a "terrible mass of fire and explosion," and "everything began to sail over their heads with the noise of falling material." As the bigger explosion occurred, *Maine* seemed to lift about two feet and the bow went down. Captain Frank Stevens, the ship's master, heard a dull, muffled explosion as though from underwater, followed at once by a terrific explosion and a dull red glare. When attempts were made to lower boats to help, some were found to have been holed by falling debris. They could hear cries from *Maine* but they didn't last long, then boats began to bring the survivors aboard, some only in their underwear or trousers. The casualty list showed that two officers and 250 men were killed and that 102 were saved, of whom eight later died. Among the dead were Sigsbee's mess attendant, Pinckney, and the Marine bugler, Newton. Most of the dead had been in the fore part of the ship.

As soon as he had recovered, Sigsbee composed a letter to the Secretary of the Navy, but, aware of the tension in Havana, he was careful to suppress the suspicions racing through his mind and urged that "public opinion should be suspended until further report." Washington found the news, which had already been doubted in Key West, hard to believe. President McKinley was stunned.

By this time *Maine* had settled, with her poop under water, and the only part of her hull visible was the torn and misshapen mass amidships and three pieces of steel jutting out of the water farther forward. The forward part of the center superstructure had been blown upward and folded back on its after part, carrying the bridge, the pilot house, the six-inch gun and the conning tower with it. The broad surface uppermost was the ceiling of the berth deck and on the white paint was the clear impression of two human bodies. The great pile of steel was so twisted that identification of details was barely possible. The foremast had disappeared and even the mooring buoy had sunk. Cellulose was still burning and ships had to shift their berths to avoid the exploding ammunition. There were many dead still in the wreck, and more kept drifting ashore.

Spanish officers and officials expressed their sympathy and offered help, and many attended the first funeral service, for 19 men, held on February 17. On that day Sigsbee put into writing what everybody had been suspecting. He cabled to John D. Long, Secretary of the Navy, "Probably the *Maine* destroyed by mine, perhaps by accident. I surmise that her berth was planted previous to her arrival, perhaps long ago. I can only surmise this."

At that time, apart from the war fever that existed, he had no evidence whatsoever for such a statement and hostility was immediately aroused. When an attempt was made to send Cuban divers down, the Spanish insisted that they must be accompanied by a Spanish diver, and that Spanish divers should be accompanied by an American. They felt that their honor was at stake, but, while a few hotheads almost hoped that a mine *had* caused the explosion, most of them were desperately afraid of war with the States.

By now *Maine,* sunk in five to six fathoms of water, was settling more into the mud each day, until eventually the poop deck was four feet under water. Bodies continued to appear and float close to the Machina. By February 21, 143 had been recovered. By this time, too, the Spanish had seen the New York newspapers. The Dreyfus case, which was tearing France apart, was filling the main news pages, but *Maine* hit the newspapers at the same time as the result of Zola's *J'Accuse* trial. There was no doubt as to which was more important. Hearst had cleared the front page of the *Journal,* saying there was no other news but that of *Maine.* "This means war," he had said, and his coverage of the disaster still stands, in the words of his biographer, as the acme of ruthless, truthless newspaper jingoism. While intelligent Americans accepted the preposterousness of the idea that Spain had blown up *Maine* and that, if she had been sunk by plotters, those with most to gain were the Cubans anyway, Hearst's newspapers refused to permit the view. "The Warship *Maine* was Split in Two by an Enemy's Secret Infernal Machine," the *Journal* said two days after the event, and even issued a seven-column drawing of the ship anchored over mines with wires leading to the Spanish fortress ashore. The following day it stated, "The Whole Country Thrills with the War Fever," and three days later, "Havana Populace Insults the Memory of the *Maine* Victims," with two days later again, "The *Maine* Was Destroyed by Treachery."

There was little doubt that the Americans were more than eager to take on the Spanish and could see the profits to be made. Senator Thurston, of Nebraska, even voiced the coldly cynical opinion that "war with Spain would increase the business and earnings of every American railroad, it would increase the output of every American factory, it would stimulate every branch

of industry and domestic commerce." Long, Secretary of the
Navy, preferred not to rush to judgment and was inclined to believe that an accident was the cause. The *Washington Evening
Star* made a point of questioning naval officers, most of whom
attributed the loss to an accident, though some felt a mine or a
bomb smuggled aboard might have been the cause. The naval
chiefs were quick to say the ship's design was sound. Rear Admiral Royal B. Bradford, Chief of the Bureau of Equipment, insisted that the coal was of good quality, and Engineer-in-Chief
George W. Melville suspected a magazine explosion. Lieutenant Frank F. Fletcher wrote to Lieutenant Albert Gleaves, of
the torpedo boat *Cushing,* that the general belief was that the
disaster had been caused by the position of the magazine next to
the coal bunkers where spontaneous combustion must have occurred. Philip R. Alger, the navy's leading ordnance expert, in
an interview with the *Washington Evening Star,* insisted that
they knew of no torpedo or mine which could possibly have exploded the ship's magazine, although magazine explosions produced exactly the effects produced by the explosion on board
Maine.

Theodore Roosevelt, however, was coming to the conclusion
that it had not been an accident. He felt that Alger was taking
the Spanish side and could not rid himself of the thought that the
ship had been the victim of an "act of dirty treachery." Always
an advocate of vigorous action, he maintained the attitude that
anyone who disagreed with him was timorous, weak and wrong.

It was during this period of inflamed feelings, on the 18th, that
Vizcaya arrived in New York. Her commander, Captain Antonio Eulate, was shocked enough to order his colors half-masted
at once, but, in view of public hysteria stirred up by the *New
York World*'s insistence that the ship was there for no other
reason than to shell Harlem and Brooklyn, tugs constantly pa-
trolled the water around her. John P. Holland, who had recently
completed successful surface tests of his newly invented submarine, was informed that the submarine must not attack *Vizcaya.* The cruiser left for Havana on February 24 without incident.

This stirring up of passions might have been excused if it had sprung from genuine patriotism, but Hearst's interest was solely in circulation and making money, and he spared nothing and no one in his vituperative lashing of the nation into a war frenzy. Nor was he alone. The *New York World* got hold of Sigsbee's suspicion that Spain was guilty of the disaster, and there was a report that a Dr. E. C. Pendleton, just arrived in Havana, had overheard a plot to blow up the ship.

Though the American press, which was not censored, gave more than the news, the Spanish press gave rather less, and by March 1 the atmosphere at Havana was volcanic with the threat of war. The Spanish cruisers *Vizcaya* and *Almirante Oquendo* arrived to a tumultuous reception, but to make matters even the American cruiser *Montgomery* arrived on March 9 to replace *Maine*. Already convinced that the explosion had come from outside the ship, not inside it, Sigsbee suggested to the Spanish that *Montgomery* should use a buoy vacated by the United States dispatch steamer *Fern* for the simple reason that it had recently been used by *Alfonso XII* and he felt there couldn't be a mine under it. The suggestion was politely rejected.

Nevertheless, Sigsbee transferred to *Montgomery* and there was an alarm the first night when distinct tapping sounds were heard, numbering 240 to the minute, a multiple of 60 and therefore considered to be from some sort of clockwork. Fearing another explosion, a great deal of listening was done in the bilges and from boats, but in the end it turned out to be from a dynamo or pump aboard *Vizcaya*, whose beat was being transmitted through the water. As the situation grew more tense, *Montgomery* left on March 17 at the suggestion of Sigsbee, who then transferred to *Fern*. By this time Cubans were suggesting that *Maine* had been sunk by the Spanish because Cuba had become an embarrassment to them and a war with the States (which they knew Spain would lose) would be the most honorable way of giving it up.

Both the Spaniards and the Americans declared their intention of holding an inquiry and divers were sent down by both sides to examine the wreck. The investigators examined quite separate witnesses, each group determined that the other should not present a one-sided report. The result was inevitable. Because each was trying to find something that would incriminate the other and avoid finding anything that would incriminate themselves, neither side was willing to accept any evidence that did not fit the conclusions each had already reached.

The Spanish inquiry started while *Maine* was still burning, but the Spanish seemed to have reached their decision even before the court, under Captain Pedro del Peral y Caballero, sat. A cordon of boats was immediately placed around the ship and security was so tight that even Sigsbee—who, having lost his uniforms, was in civilian clothes and was not identified—was not allowed to approach.

On February 20, only five days after the disaster, the Spanish took the strange step of announcing a preliminary verdict. The wreckage had not yet been examined and no witnesses had been interrogated, but, fearing a hasty action by the Americans, they wanted to show that such an action would be unjustified. When the full report appeared, it merely confirmed the first premature opinion. The explosion, the Spanish court decided, could not have been caused by a mine. In those days mines—or torpedoes, as they were often called—were of two kinds: contact mines set off by brushing against a ship, or electrical, fired from the shore. But no trace of cable had been found in the harbor. There was also no instance, they pointed out, of a mine's exploding a ship's magazine, and yet it seemed certain that one, possibly two, of *Maine*'s forward magazines had blown up.

The Spanish confined their examination to the neighborhood of the wreck without examining the wreck itself very much, and they laid themselves open to criticism by their hasty preliminary report. Nor did they properly question the survivors, though they did apply to the Americans for the facilities and were refused. They did not press the point, probably realizing there would be difficulties, and the court did not deal with questions which were already being asked in the American press. Had mines been laid in Havana harbor? Were the authorities in Cuba in possession of explosives or torpedoes of a kind that might have accounted for

Maine? With war imminent, it was not unreasonable for these points to be ignored, but this failure certainly prejudiced Spain's case in the eyes of the world.

In short, the report stated simply that *Maine* could not have been sunk by an exterior explosion, and that therefore the cause must have been spontaneous combusion in the coal bunkers or something of that nature. They pointed out that Don Saturnino Montojo, "an illustrious lieutenant in our navy," had mentioned the case of *Reina Regente,* a Spanish warship built on the Clyde, in which an explosion had occurred while she was still in the hands of the builders. This was traced to the formation of gases in a watertight compartment, and the court suggested that something similar might have happened to *Maine*'s bunkers. They also pointed to the evidence of Lieutenant Peres y Perera, who had witnessed the whole thing but had seen no column of water such as a mine would have produced. No shock had been felt ashore, they pointed out, and there were no injuries to the piles of the nearby piers and no dead fish in the harbor. Whenever small blasts had been discharged in the bay to loosen rocks or remove the hulls of sunken ships there had always been dead fish. Spanish divers had also reported that there was no fissure or other shaping of mud near the wrecked ship, such as might have been expected from an external explosion. Therefore, the explosion must have been caused originally by an explosion close to the forward magazine which destroyed decks and bulkheads.

The theory of spontaneous combustion in a coal bunker was not farfetched because there had been several similar cases in American and other foreign ships, one or two, in fact, just before the disaster to *Maine.* There was also a bunker fire in *New York* soon afterwards, and one point foreign experts noticed was that some of *Maine*'s coal bunkers, of which there were 20, ten on each side, were very close to the magazines.

Nevertheless, the calm detachment of the Spanish infuriated the Americans who already suspected sabotage. The American court met at Havana on February 21, six days after the disaster and the day after the Spanish court's preliminary report. This inquiry lasted a month and was much more thorough than that of the Spanish. But, like the Spanish, it could not probe too closely into the other country's secrets. If it would have been an offense for Spanish divers to probe into the wreck because it still retained the extraterritorial rights of a man-of-war, in the same way the Americans could hardly have dragged the harbor for traces of electric cable or cross-examined Spaniards about their supplies of explosives or the placing of floating mines. While the Spanish hunted the harbor for clues they were not too anxious to find, the Americans searched for signs of an interior explosion which most of them already believed to be too much of a coincidence to be possible. Obviously, there should have been an international court, but in 1898 the machinery for settling international disputes was very rudimentary and, in any case, neither side would have accepted a verdict which might reflect on national honor.

The American court took evidence from the officers and crew of *Maine,* who reported on the state of the ship before the explosion, and their experiences on the night of February 15. In careful statements, Sigsbee set forth his position. The Spanish knew he was coming (so he understood); the buoy was seldom used (he was later told); the coal taken aboard at Key West was inspected (though he could not remember the event, such was the custom); he did not know how much coal was in the bunkers (but assumed there was not much); he knew the fire alarms worked (because they occasionally went off at temperatures below that at which they were set); to the best of his knowledge, all regulations concerning the stowage of inflammables and paint and the disposal of waste and ashes were strictly enforced (though he had not seen this, he had given orders); and he was certain he had inspected the magazines during the last three months (though he could not be precise as to the date).

The testimony showed a man who, while a brave seaman, was unfamiliar with his ship, might not have understood the complexities of a modern vessel, and probably suffered from the division existent in all navies of that time which separated deck officers from engineers. His vagueness probably also stemmed from the belief that giving an order meant it was executed. Whatever the reasons, Sigsbee appeared to have been isolated from day-to-day routine.

Following the officers, evidence was heard from the experts and divers who examined the wreck and the harbor bottom nearby after the disaster. Conditions were poor and visibility was bad, and soft ooze had hampered walking on the harbor bottom. Pieces of torn wreckage had not only been difficult to identify but were also a danger to air hoses, and many of the navy divers were not highly professional.

On one point both courts were agreed: there had been two explosions, the original one, whatever its cause, and the explosion of the magazines which followed. There was always the possibility that a bomb or an explosive device had been smuggled aboard, but visitors had always been followed and closely watched. The theory of spontaneous combustion in a coal bunker offered by the Spanish had also been carefully considered. In *Maine* there were six bunkers directly abutting the magazines, but four of these, it was claimed, were empty and had just been painted, and a fifth was only half-full and had been in use on the evening of the disaster. The remaining bunker, holding 40 tons of soft New River coal, might well have caused the trouble, but the coal had been carefully inspected before being taken aboard, and again by the duty engineer officer only 12 hours before the explosion. Three of the bulkheads surrounding it, it was stated, were easily accessible and would probably have been touched by anyone going down the narrow passage to the loading, hydraulic and dynamo rooms, and also by crew members who were in the habit of lounging against it. The fourth bulkhead had just been painted. A rise in temperature, the danger signal for spontaneous combustion, would surely have been noticed. In addition, all bunkers were equipped with electrical alarms and thermometers, and the temperatures regularly taken and recorded.

No smokeless powder and no explosive shells which might have erupted were stored in any of the magazines, and the torpedo heads, primers and detonators were all stored aft under the wardroom and were unaffected. There were no steam pipes or dangerous electric wires in the magazines, the lights being boxed in and wires and lights alike separated from the magazines by a double plating of glass. No loose powder was allowed to be about, the magazines were kept locked and proper soft shoes unlikely to cause sparks were worn by men working in them. As for the suggestion of an explosion arising from gases, with ventilation pipes and thermometers and a proper system of inspection, the danger should have been negligible. Only two boilers were in use at the time, and these were remote from the magazines and in good condition and, with the ship lying in harbor, were working at low pressure anyway. Finally, there had been no carelessness in handling inflammable materials because waste was carefully looked after and varnishes, driers, alcohol and other such combustibles were stowed on or above the main deck.

By eliminating all these possibilities, the American court, which issued its report on March 21, concluded that the explosion had been caused by some agency outside the ship, though they did not think this was a torpedo. It would have been impossible for a torpedo tube to have been mounted ashore and discharged without the connivance of the authorities, but since the ship lay within 800 yards of the shore, it was not beyond possibility that *Maine had* somehow been torpedoed.

The Americans had carefully examined the wreck of the hull, however. Ensign Wilfred V. N. Powelsen, of *Fern,* who had studied at Glasgow with a view to becoming a naval constructor, was in charge of the divers, and it was discovered from his efforts that at Frame 17, the outer shell of the ship, from a point eleven and a half feet from the middle line of the ship and six feet above the keel when in its normal position, had been forced up to about four feet above the surface of the water, in other words about 34 feet above where it would have been had the ship sunk uninjured. The outside bottom plating was bent into a reversed V-shape, the after wing of which was doubled back on itself. At Frame 18 the vertical keel was broken in two and the flat keel bent into an angle similar to that formed by the outside bottom plating.

The explosion, Powelsen felt, could not have taken place internally because the forward six-inch magazine, where it would have occurred, and the equipment stored near the magazine showed no signs of burning. He carefully dispensed in the same way with the possibility of explosions in other magazines. He had found the starboard side of the ship blown outward, and because the ship had been pushed violently from port to starboard, he had come to the conclusion that there had been a heavy external explosion on the port side.

This effect, the court felt, could have been produced only by the explosion of a mine situated beneath the ship, which caused the partial explosion of two or more of the forward magazines, but it declined to fix responsibility for the placing of the mine. The verdict confirmed what Sigsbee had suspected all along, and had been anticipated long since by Hearst's newspaper, which stated quite falsely that "the Court of Enquiry finds that Spanish government officials blew up the *Maine*." The paper claimed that the warship "was purposely moved where a Spanish mine exploded by Spanish officers would destroy it," and staked its reputation on the certainty of war.

Though Sigsbee, in a letter to his wife, said he feared nothing which would reflect on his handling of *Maine,* many Europeans were unconvinced by the American report. On the Continent it was boldly asserted that the American court of inquiry was prejudiced and had made its mind up before it started to take evidence, and that *Maine* had been sunk through neglect, carelessness or bad discipline. Some of these charges, however, could well have sprung from that same dislike of the young and powerful United States engendered among the older and fading European powers that had bothered Flood in Gibraltar 25 years before. Opinion was about even between those who believed the Spanish had blown up *Maine* and those who believed she had blown herself up.

In an attempt to calm ruffled feelings, President McKinley pointed out that *Maine*'s arrival in Havana had caused no appreciable excitement; on the contrary, there had been relief at the resumption of friendly intercourse after the recent tension. The Consul General also tried to cool down tempers by saying there was no suggestion in Havana that there should be war and urging that the presence of American ships should be maintained. Lee's view was that a mine had caused the disaster but that it had been an accident and that the Spanish government was innocent of complicity. But while fair-minded men pleaded for calm and one paper at least pointed out the absence of technically qualified members at the court of inquiry, other newspapers continued to stir up the war fever. Hearst had earlier telegraphed one of his artists in Cuba who had felt there would be no war, saying, "You furnish the pictures. I'll furnish the war." It was obvious that Spain, proud but grievously in debt, would never want a war with the United States and had been going to enormous lengths to swallow the insults in the American press. But that did not stop the American press, especially the Hearst papers, from inventing murders and atrocities which intelligent Americans knew could not be true, while the Cubans, naturally more than interested in drawing the United States into their quarrel, had added their own stories of rape, torture and murder. Though the papers spoke for the American nation, the nation had little idea of what was really going on in Cuba. It was little wonder that Cánovas del Castillo, the assassinated Prime Minister of Spain, had once remarked to an American correspondent, "The newspapers of your country seem to be more powerful than the government."

The arguments went on. Rear Admiral Bradford, who claimed to have witnessed more underwater explosions than most naval officers, told the Senate Committee on Foreign Relations that he had no doubt that a mine had been the cause. Sigsbee's evidence to the committee was as vague as ever. He had attended only some sessions of the court of inquiry, he said, and preferred to be judged by the people. It was his opinion that *Maine* had been destroyed by a mine and he was still convinced that a party of men had laid one made out of an old hogshead or wine cask. How else, with his precautions, could it have happened? He hastened to point out that the ship, during all her time in Havana, had never before swung in the direction she did that night— into a position where her guns could have put the Spanish batteries under fire. If he had been given the task of defending the port with a single mine, he said, he would have chosen that spot to lay it. He even hinted at information he possessed which he could not disclose, but no one thought to question him on the point, and he drew plans showing how a mine could have been placed where *Maine* would swing if she took up a position dangerous to the port, and how a mine could have been planted beneath her by a lighter coming alongside trailing electric cables to the shore. He pointed out that, despite what President McKinley had said, *Maine* was not welcome in Havana and had been taken to a mooring buoy which seemed to be reserved for some special purpose not known to the pilot.

Ignoring the repeated assurances of the Spanish government that they knew nothing about the explosion and that their military authorities were not responsible, the Americans felt they were quite justified in going to war. There was no question of further investigations. By then it was too late, anyway, and America—and certainly William Randolph Hearst—was glad of the opportunity of clearing this European neighbor from her doorstep. A month later Congress recognized the independence of Cuba and directed McKinley to take steps to compel the Spanish to evacuate the island. The Spanish Minister in Washington called for his passports and the war began.

It was an unnecessary war, a newspaper's war. Above all, it was Hearst's war. Had it not been for his tawdry flair for publicity, there would probably have been no war at all, and when it came he was "in a state of proud ecstasy," accustomed to refer to it as "our war." At one point even, to the disgust of his staff, he put up the headline "How Do You Like the *Journal*'s War?" In addition to hostilities, of course, he had achieved a circulation of a million copies a day.

Vacant stores on every avenue in New York blossomed with recruiting stations, with pictures of *Maine* before and after the explosion, while men from the *Social Register* waited at the door to pull recruits in. Those who recruited enough could expect to be made officers in the regiment by the man who had rented the store, who intended to be the colonel, while the men who enlisted "could qualify for a drink in almost any bar between Eighth and 23rd Streets, and between the Hudson and East

Rivers," together with "a red flannel bellyband from any patriotic young female."

"Remember the *Maine*" became the battle cry of troops going into action, but the navy denied a newspaper story that they had flown the cry as a signal in their two sea battles with the Spanish navy. After the Spanish surrender of Santiago, Hearst correspondents plastered the city with posters reading, "Remember the *Maine*," with "Buy the *Journal*" at the bottom. The American commander, General W. R. Shafter, feeling that their action was an open invitation to a massacre of Spaniards by vengeful Cubans, had the Hearst men arrested and sent home.

The war ended Spanish domination of Cuba and increased the United States' prestige. It seemed splendid for those Americans who were far from battle, but for the soldiers it was as grim, dirty and bloody as any other war, and only the incredible ineptitude of the Spanish and the phenomenal luck of the Americans prevented it from stretching into a struggle as long and disastrous as the Boer War was for Britain. Deaths in battle were few, but revelations of corruption, the inadequacies of the War Department, quarrels between rival commanders and the imperial spoils with which the United States emerged gave the war a disreputable look.

Hearst took his private yacht to Siboney armed with a printing press with which he printed a newspaper for the troops. As Charles Johnson Post, one of his cartoonists, who rushed off to join the colors, said, "It hadn't much news, but paper was scarce in the trenches and it was welcome." Since army stationery was also soon exhausted, orders were frequently written on its wide margins. Hearst later distributed medals made from metal taken from *Maine*.

Maine had produced the war the Americans had been wanting and expecting. It had been brewing all the time, despite Spain's desperate efforts to avert it, and *Maine* was merely the trigger that set it off. But had the truth really been reached?

Sigsbee was never in any doubt. "If an expert," he wrote, "had been charged with mining the *Maine*'s mooring berth, purely as a measure of harbor defense, and having only one mine available

. . . he would have placed it under the position the *Maine* occupied that night." He never swerved from this view and reiterated it in the *Century Magazine* in 1899, his chief reasons being simply that Spain was unfriendly and that *Maine* was not welcome. Once more he explained how it could have been done, with a mine slung under a lighter and released as the lighter approached *Maine*.

His attitude would be the natural reaction of a man trying to avoid any suggestion of neglect of duties, and a serious attack on his theory came from Lieutenant Colonel John T. Bucknill, of the Royal Engineers, in the pages of *Engineering*, a British professional journal. Bucknill had been secretary of a joint Admiralty–War Office committee which had carried out experiments with explosives against the double bottom of H.M.S. *Oberon* between 1874 and 1876, and he concluded that the findings of the American court of inquiry were absurd. He pointed out that since Havana harbor had a long, narrow entrance admirably suited to defense by mines, it was incredible to suggest that the best way to defend it was to plant the mines *inside* the harbor. He refused to believe that the mine could have been laid within 400 yards of the German training ship *Gneisenau* without being seen, and even more so that it had been laid after *Maine* had arrived. The damage to *Maine*, Bucknill wrote, was not in accordance with the effects of a mine explosion. He decided that spontaneous combustion in the coal bunker caused the first explosion in or near the six-inch reserve magazine and that the second was caused by some of the slow-burning powder in the 10-inch magazine, and it was the second explosion that had done the damage.

Rear Admiral George W. Melville, Chief of the Bureau of Steam Engineering, also attacked the mine hypothesis, in 1898 and again in 1911. He pointed out that Spain did not want war and was even struggling to avoid it, as demonstrated by her sending *Vizcaya* to New York, where she would have been virtually in the hands of the enemy. He also pointed out that no one had ever come forward to accuse any individual, to claim a part in a conspiracy or to say he was a member of the working party which had laid the mine.

Though in Europe it was generally believed that *Maine* had not been destroyed by the Spanish, the Americans remained satisfied that justice had been done, and there was no further investigation until 1909, long after the war was over, when the battleship's wreckage was declared a danger to navigation. Sigsbee and Captain French E. Chadwick, a member of the original court of inquiry, were convinced that an examination of the wreckage would confirm their beliefs, especially as during the Russo-Japanese War Russian ships had been lost to mines which had detonated their magazines. What they did not realize, however, was that the Russians were using a volatile and unstable smokeless powder while *Maine* was carrying the more stable black and brown powders.

In 1910 Congress finally voted half a million dollars for the project and the work was entrusted to the army engineers. Since *Maine* had become deeply embedded in the mud by this time, it was decided to enclose the wreckage in a huge cofferdam of steel piles driven through 20 feet of mud into the solid clay beneath and then pump out the enclosed space. A year later, however, the work was still unfinished and more than double the amount of money voted by Congress had been spent. The plan was therefore modified and the after part of the ship, which was reasonably intact, was separated from the wreckage, bulkheaded, floated, towed away and sunk in deep water. A smaller cofferdam built around the forward portion of the wreckage was completed in 1911.

The wreckage was then carefully examined by a naval constructor, Mr. W. B. Ferguson, whose conclusions tallied with those of the original American court. His chief discovery was a curved bottom plate, which, he maintained, like Ensign Powelsen, indicated an exterior explosion by a mine. This mine, he said, filled with a low explosive that would distribute the explosion over a wide area, came in contact with *Maine*'s hull at a point below the six-inch magazine where the black powder used for salutes was stored. This was ignited and fired the other forward magazine.

This final report, though accepted by the navy, was by no

means accepted universally. Bucknill was no more satisfied with it than he had been with the original findings. "Nothing," he said, "warranted the theory of an external explosion," and he pointed out that, despite the lapse of 14 years, no one in Havana had yet stepped forward with even a whisper of evidence that a mine had been planted in the inner harbor.

The belief in Spain's innocence rested to a certain extent on a report published in British newspapers in 1911, before any examination had been made, which claimed that Brigadier General William H. Bixby, the engineer in charge of the operation, had said that the magazines could not have been exploded from outside the ship and that indications on the hull proved that the explosion had come from inside. He was also reported to have said that the cause could not have been a Spanish mine and that "a terrible mistake had been occasioned."

This statement was allowed to remain unchallenged for nearly two years and attracted more notice in Europe than the official report. When at length an article in the *Fortnightly Review* referring to it was brought before Bixby's notice, he denied he had ever said anything of the kind, and pointed out that the examination of the wreckage was outside his province and that, anyway, when the reports first appeared, the wreck was still buried in the mud.

It was accepted then, and has been since, largely because of the writings of J. G. Lockhart, that *Maine* was destroyed by a Spanish mine, but not necessarily with the knowledge of the Spanish authorities. It came to be believed that when *Maine*'s visit could not be prevented, in view of the strained relations the reasonable precaution of laying an electrically controlled mine at the spot where *Maine* was to be moored was taken, or at least that it was arranged for her to lie at a spot where such a mine already existed. If war had broken out, the Spanish would then have been quite justified in destroying *Maine,* just as *Maine* would have been justified in turning her guns on the city and the forts. The anchorage, Lockhart said, was determined by the port authorities and a Spanish pilot brought *Maine* to her buoy, which, it was noticed, was not in the place marked on the chart. During the evening, *Maine* swung, as Sigsbee noticed, to the position she would have taken up to engage the batteries on shore, and if there was a mine, Lockhart felt, it was doubtless laid with this possibility in view. The Spaniards, he continued, must have known the truth and should have made a clean breast of it, but they did not, and the United States was left with no choice but to go to war.

Prejudice continued well after the incident was formally closed, and even today the view is prevalent that because of the desire for a war with Spain, the Americans ignored evidence that would have exonerated the Spaniards. And there is some justification for the belief. A mine produces a heavy, dull concussion on exploding, whereas most of the witnesses described the first explosion as a sharp report like a gunshot. A mine would also have thrown up a column of water, but none was seen—though it could have been obscured by the smoke which poured immediately from the ship. The mine could, of course, have gone off directly beneath the hull, but that would have thrown the ship to one side or lifted her, and most of the witnesses mentioned only the upheaval following the second explosion. No concussion was felt by other ships save the steamer *Deva* lying 600 yards away, whose captain thought he had been collided with. And the Spanish comment on dead fish was rejected by the Americans, who suggested somewhat naïvely that the fish were probably in the habit of leaving the harbor at night or, alternatively, that the explosion had only stunned them and that they later recovered and swam away.

A more serious objection to the American findings remained. During the First World War when mines were used on a far greater scale than previously, there was only one occasion when a mine exploded a ship's magazine. This was in H.M.S. *Russell* off Malta in 1916, and then only the shell room was exploded. Though it could sink a ship, a mine was unlikely to explode its magazine, a fact the American court did not even consider.

Since the Spaniards would hardly have laid a contact mine in Havana harbor which was frequented by ships of all nations, including their own, any mine that destroyed *Maine* must have been electrically controlled from the shore, in which case the

Spanish were guilty of treachery. At the time, they denied that
any mines had been laid off the island's coast, but early in April
after the explosion it became known that they did have mines in
Cuba and were using them. Vice-Admiral Beranger, Secretary
of the Spanish Navy before the war, told a reporter from Madrid's
leading evening paper, *El Heraldo,* on April 6 that attacks on
Cuban ports were not expected because "Havana, as well as Cien-
fuegos, Nuevitas and Santiago, are defended by electrical and
automobile torpedoes." Beranger added that on August 8, 1897,
six months before the explosion, Cánovas del Castillo, the assassinated premier, had arranged with him to have 190 mines installed in these ports by an expert.

It could also be pointed out that the Spanish took great care
to insist that *Vizcaya* should be protected in New York, and it
was at their insistence that tugs were constantly patrolling
around her. Knowing they had a mine near *Maine,* were they
afraid that the Americans might try to retaliate with something
similar?

But assuming there *was* a mine, why explode it? It doesn't
make sense to suggest that the Spanish authorities were responsible because it would not only have been an act of treachery
that would have blackened them in the eyes of the world, it would
have been an act of supreme folly. It would have made war with
the United States inevitable and they could never have wanted
that, with American bases just across the Straits of Florida and
their own main bases 3000 miles away across the Atlantic. Perhaps, however, so the belief ran, some subordinate obtained access to the electrical gear controlling the mine, and the tempta-
tion was more than he could resist.

A later view, more sympathetic to Spain, suggested that the
mine was there all right but had been set off by a Cuban who,
knowing that America already sympathized with the Cubans,
guessed that if *Maine* were damaged by a Spanish mine, it would
bring the Americans in on their side, and had pressed the plunger.
Ferdinand Lundberg, in his antagonistic book on Hearst, suggested even that Hearst himself might have had some connection
with the explosion. When Hearst died in 1951 the *Manchester*

Guardian said of him, "No man has ever done so much to debase the standards of journalism," and certainly his pronouncements and his pride at having "furnished" the war did not make it entirely impossible that he or his money had something to do with it.

These were all possibilities until 1968, when they were all firmly rejected by Mr. K. C. Barnaby, OBE, an honorary vice-president of the Royal Institution of Naval Architects and a leading naval architect with long experience of ships and the sea. He came down quite firmly on the side of an internal explosion.

The cloud of smoke and fire seen to rise from *Maine* was a symptom of magazine explosions, he wrote, and he felt that the flash traveled far aft under the armor deck, although only the forward magazines were involved. He pointed out that explosives deteriorate and become dangerous if kept too long, and drew attention to the Japanese battleship *Mikasa*, which was sunk in 1905 by the spontaneous ignition of spoiled explosives. In 1906 the Brazilian battleship *Aquidaban* was also destroyed by an explosion that was almost certainly due to the decomposition of her smokeless ammunition. She was stationed in a hot and humid climate like *Maine*, the sort that was apt to make nitroglycerin propellants dangerous if kept too long. The American report, though it described the precautions taken with *Maine*'s explosives, did not mention how old they were.

Certainly disasters caused by unstable explosives had so far evoked little alarm among the larger navies, but the situation was suddenly altered by the explosion in the French battleship *Iéna* in 1907 as she lay in dock in Toulon. The first explosion, as with *Maine*, was not severe but was followed at once by a second of extreme violence. It was traced to "B" powder, a smokeless nitrocellulose powder made up in cartridges containing a small quantity of black powder to detonate the smokeless main charge —mentioned in the postwar report on *Maine* by Ferguson. This nitrocellulose and black powder mixture was afterwards found to be a very dangerous combination. In *Iéna* one of the magazines was in a hot position near the dynamo flat, while the ventilation was under repair.

Following this disaster, great quantities of old explosives were destroyed by dumping or burning. In 1908, however, the decomposition of old explosives destroyed the Japanese cruiser *Matsushima,* and in 1911 explosions in the battleship *Liberté* showed that the French navy was still retaining explosives far too long. The ammunition for this explosion once again contained "B" powder.

According to Barnaby, even Powelsen's evidence was doubtful. The curiously inverted V-shape of the keelplates found in the wreckage—one of the strongest arguments for a mine—was subsequently found to be an effect that could be produced by an internal explosion. Powelsen had also noted that the starboard side of the ship had been blown out and this, curiously, helped convince him there had been an external explosion. Barnaby didn't agree and felt that the detonation in *Maine* was of the same type as those which wrecked so many other warships in later years and was not the result of a mine.

Experts have a depressing habit of taking the romance out of mysteries, and perhaps the last word on the *Maine* disaster has been said by Admiral H. G. Rickover, of the United States Navy. Rickover, the U.S. Navy's atomic expert and the man behind their atomic submarine program, came across the fact, not generally reported, that following the war against Spain, Sigsbee was given the battleship *Texas* and once more his ship was found to be dirty and rusty and its boats not properly outfitted.

From his reexamination of the records, Admiral Rickover concluded firmly that there was no evidence whatsoever that a mine had destroyed *Maine* and that the disaster had been caused by a fire in a bunker. Such fires had occurred before in bituminous coal such as *Maine* carried. They were difficult to detect and often smoldered deep below the exposed surface of the coal, giving neither smoke nor flames nor raising the temperature in the vicinity of the alarm. The bunker of *Maine* had not been inspected for nearly 12 hours before the explosion, a period, he said, which experience had shown was ample time for a fire to begin, heat bulkheads and ignite the contents of adjacent compartments.

The President of the United States, Rickover claimed, was unfortunate in *Maine*'s commanding officer. Although Sigsbee took the proper precautions to protect his ship against harm from external sources, there is no evidence that he took more than routine measures to safeguard her from accident. He knew of the danger of spontaneous combustion of coal and must have been aware of bunker fires in other ships. The fact that the bunker alarms sounded below the danger point was no reason for feeling safe because they were inaccurate, and, as Rickover comments, the significant point was that under his command both *Kearsage* and *Texas* were found to be dirty. Sigsbee's evidence reflected the times: the word of a senior officer was accepted without question.

Rickover's view was that the courts of inquiry had been deficient in their duty, and that Sigsbee's testimony and later writings on the subject were vague and speculative. The court's verdict was one that could be expected, however, because of the strained relations between the United States and Spain, the warlike atmosphere in Congress and the press and the natural tendency to look for causes that did not reflect adversely upon the navy. Had the ship blown up in an American or a friendly port, Rickover felt, it was doubtful whether an inquiry would have laid blame on a mine. The finding of the court of 1898 appeared to have been guided less by technical consideration than by the awareness that war was inevitable.

When the Senate Committee on Foreign Relations investigated the findings, Rickover continued, it produced nothing of technical substance and seemed to exist merely to provide a record of Spanish misdeeds. Officers who had not even seen the findings of the court were asked for their opinions, and Sigsbee was allowed to relate his theories of a makeshift mine without challenge. As for the board of 1911, doing its work free from the risk of war and coming to the same conclusion as the board of 1898, according to Rickover, it reached its findings without the advice of any outside experts and without the help of any available technical information. He pointed out that only 13 years had elapsed since the nation had gone to war with "Remember the *Maine*" as a battle cry and it would have been difficult to suggest that the constituted authorities had made a grave error.

By contrast, he said, the French investigations into the magazine explosion in *Iéna* in 1907 were particularly exhaustive, but, of course, the explosion this time occurred in a home port and at a time when the international situation was calm.

Maine, he said, was without doubt destroyed by an explosion inside the ship, but he pointed out that whatever President McKinley felt or did he risked war. If he did nothing, public sentiment might have forced it on him, and if he exerted pressure on Spain he could also have found himself in arms against Spain.

Rickover's theory is supported by similar disasters in the 1914–18 war. Many were attributed to German agents under the legendary von Rintelen, but now it seems clear that they were caused in the ships' magazines. One author has even suggested that the destruction of *Royal Oak* in 1939 was caused by an internal explosion and not by Gunther Prien's torpedoes.

As a final check, Admiral Rickover persuaded two more experts, Mr. Ib S. Hansen, Assistant for Design Applications in the Structures Department at the David W. Taylor Naval Ship Research and Development Center, and Robert S. Price, a research physicist for the Naval Surface Weapons Center, to make an investigation. They had no doubt that the courts of inquiry were wrong in their findings, and pointed out that between 1894 and 1908, more than 20 coal bunker fires were reported in United States naval ships. From the fact that extra bulkheads were installed to surround the magazines, it was clear that a number of people in the navy did not believe that the single bulkhead system in *Maine* was safe, and they felt that the reason for the magazine explosion was probably heat from the coal bunker adjacent to the six-inch reserve magazine.

Whatever the cause, Rickover decided, the chances of peace died with the destruction of *Maine.* Congress and public demanded intervention and it was under these circumstances that the first court of inquiry did its work. Had they approached the matter under different circumstances, the court might have reached an entirely different verdict and the result would have been an injection of reason into an emotional atmosphere, so that the United States would not have found itself adopting an attitude which was technically unsound and which has been increasingly questioned over the years.

Was it possible that someone did realize that *Maine* had been destroyed by an internal explosion but, set on war and carried away by public opinion, suppressed what had been discovered? The explosion, though an accident, provided a splendid reason for hostilities, and the theory would make sense of that statement by General Bixby, later quashed, that there had been "a terrible mistake." Had he unguardedly admitted the mistake and then, faced point-blank with what he had said, been obliged to deny it? After two years, he had had time to think or to have it pointed out to him that, since America had gone to war over *Maine* and hundreds of men had died on both sides, it was a dangerous thing to admit. Though Bixby stressed that his statement had been made before the wreckage was examined, he was a senior officer and could easily have had access to probable findings. This seems even to be borne out by President McKinley's first reaction. (McKinley, though an enigmatic man and not a distinguished President, was always highly regarded for his honesty and straightforwardness.) Impressed by the conciliatory efforts of the Spanish government, McKinley made great efforts to cool a dangerous situation; he was obviously not assuming that the Spanish had attacked.

Rickover's view seems sound. Another American admiral said of his opinion, "Written by a man of Rickover's reputation, it would be enough for me." America went to war over a *casus belli* that didn't exist, but, as it happened, the sinking of *Maine* did not create the emotional forces that existed; it merely released them, and the result was that the United States became an imperial power.

4. S.S. Waratah (†1909)

The liner that vanished

According to Alan Villiers, ships which simply disappeared leaving no trace vanished at a rate of five to 10 a year in the 16 years before 1975, and between January 1961 and January 1971, 70 merchantmen, ranging in size from very small to more than 13,000 tons, were officially posted missing at Lloyds in London. They were all built, loaded, classed, surveyed and controlled under proper supervision, and there are no longer wreckers, pirates or coffin ships. But still they vanish.

During the 1840s and 1850s, the early years of the change from sail to steam, the number of missing ships seemed even greater, and the losses were often the result of ship designers' producing inadequate vessels. With the advance of technology, the Victorian era contained all too many men both in Britain and elsewhere who built too fast and without really knowing the science of their trades—men on stilts, as one author has called them.

The period was one when the Royal Navy, though it had long since lost the skill and initiative it had enjoyed in Nelson's day, was still resting heavily on the laurels of Trafalgar, revered out of all proportion to its then ability. As a result, there was a tendency among naval officers, from their pinnacle of privilege, to consider that they knew more about building ships than naval architects and shipbuilders. It is no coincidence that among the most vociferous supporters of the theory that naval men should design their own vessels was Admiral Sir Edward Belcher, who had lost his ships in the attempt to find Franklin. H.M.S. *Captain,* a battleship, was built to the design of an ambitious and self-important officer, and it is hardly surprising to learn that *Captain* disappeared in 1870 on only her third cruise, in a squall off Cape Finisterre which caused no problems to the other ships in her squadron.

The emigrant trade to the United States, Australia and New Zealand, starting in the early days of steam and before the arrival of Samuel Plimsoll's compulsory load line, produced many unscrupulous or even unknowing shipowners who overloaded their ships. In those days ships crossing from England to North America took a more northerly route—abandoned after the *Titanic* disaster—which carried them too near the path of icebergs coming down from the Arctic, such as were encountered by the brig *Renovation* when she saw what were thought to be Franklin's ships. Several ships were lost to gales or to the ice, among them the paddle steamer *President,* which left New York on March 11, 1841, with among her 136 passengers and crew the son of the Duke of Richmond and Tyrone Power, a famous Irish comedian, said to be an ancestor of the film star of the same name. Despite her size, *President*'s reputation was dubious after three Atlantic crossings, and her master, Captain Richard Roberts, had gone so far as to describe her as a "coffin ship." When she failed to appear, rumors abounded and the inevitable messages in bottles and supernatural visitations appeared. *President*'s seaworthiness had often been questioned, and among those who knew, there was little surprise when she vanished without trace.

City of Glasgow, a steam-sailing ship, disappeared in roughly the same area in 1854. A black derelict with a bright red bottom, believed to be she, was seen in the Atlantic by *Mary Morris,* out of Glasgow, on August 18, at a time when *City of Glasgow* should already have been up the Delaware. A ship called *Baldaur* also saw what she took to be the derelict ship, and, as she drew near, a small bark shot away from her side. There were boxes and crates in the sea and a suggestion of piracy, but it may well have been merely a "starvation ship" (a ship whose crew were

Sir John Franklin

Lady Franklin

H.M.S. *Erebus* and *Terror* parting from their escort ships in June 1845

H. M. S.hips *Erebus* and *Terror*
{ Multiedin the Ice in

28 of May 1847 } Lat. 70° 5' N Long. 98° 23' W

Having wintered in 1846—7 at Beechey Island
in Lat 74° 43' 28" N. Long 91° 39' 15" W After having
ascended Wellington Channel to Lat 77° and returned
by the West side of Cornwallis Island.

Sir John Franklin commanding the Expedition.
Commander.
All well

Party consisting of 2 Officers and 6 Men
left the Ships on Monday 24th May 1847

Gm Gore Lieut
Chas F Des Vaeux Mate

WHOEVER finds this paper is requested to forward it to the Secretary of
the Admiralty, London, *with a note of the time and place at which it was
found*: or, if more convenient, to deliver it for that purpose to the British
Consul at the nearest Port.

QUINCONQUE trouvera ce papier est prié d'y marquer le tems et lieu ou
il l'aura trouvé, et de le faire parvenir au plutot au Secretaire de l'Amirauté
Britannique à Londres.

CUALQUIERA que hallare este Papel, se le suplica de enviarlo al Secretarie
del Almirantazgo, en Londrés, con una nota del tiempo y del lugar en
donde so halló,

EEN ieder die dit Papier mogt vinden, wordt hiermede verzogt, om het
zelve, ten spoedigste, te willen zenden aan den Heer Minister van de
Marine der Nederlanden in 's Gravenhage, of wel aan den Secretaris den
Britsche Admiraliteit, te London, en daar by te voegen eene Nota,
inhoudende de tyd en de plaats alwaar dit Papier is gevonden geworden.

FINDEREN af dette Papiir ombedes, naar Leilighed gives, at sende
samme til Admiralitets Secretairen i London, eller nœrmeste Embedsmand
i Danmark, Norge, eller Sverrig. Tiden og Stœdit hvor dette er fundet
önskes venskabeligt paategnet.

WER diesen Zettel findet, wird hier-durch ersucht denselben an den
Secretair des Admiralitets in London einzusenden, mit gefälliger angabe
an welchen ort und zu welcher zeit er gefundet worden ist.

A facsimile of a message signed by officers of Franklin's expedition,
found by Captain McClintock on Prince of Wales Island

"They forged the last link with their lives"— an artist's impression of the end of the Franklin expedition

Sir Robert McClure, who saw the Northwest Passage while searching for Franklin in 1850, dressed in "Arctic clothing"

The crew of *Dei Gratia* discover the *Mary Celeste*

Mary Celeste in full sail

Captain Morehouse of *Dei Gratia*

Captain Charles Sigsbee
of the *Maine*

William Randolph Hearst,
the American newspaper
publisher who did much
to whip up war fever
against Spain

The *Maine*'s
barnacled wreck
surrounded by
a cofferdam,
1911

The Blue Anchor liner *Waratah*, built in 1908 and lost in 1909

The *Waratah's* master, Captain Ilbery,
the chief officer, Mr Owen, and the crew

The U.S. Naval Auxiliary collier *Cyclops*

Joyita as she was found, waterlogged and drifting

Joyita being towed to Suva

Donald Crowhurst leaving Teignmouth aboard *Teignmouth Electron*

Teignmouth Electron being towed into Santo Domingo by the freighter *Picardy*

getting short shrift at mealtimes for a variety of reasons, such as a miserly owner or a dishonest cook, steward or even captain), trying to pick up supplies from the abandoned vessel.

In the same year *Lady Nugent,* a troopship with nearly 400 men, women and children aboard, disappeared in the Indian Ocean. Two years later it was *Pacific,* a brand-new Collins liner, which was believed to have struck an iceberg with a loss of 200 lives. The following year *Tempête* disappeared, then from 1869 to 1873, *United Kingdom, City of Boston,* an Inman liner and *Ismailia* were added to the list. The Atlantic was full of derelicts. Over 900 floating wrecks, each a danger to shipping, were sighted between 1887 and 1891, many of them in the North Atlantic. A large percentage of them had capsized and only a third could be identified, and a special vessel was chartered to sink them.

But these ships had all vanished before the turn of the century. By 1900 it was felt that steam had been mastered and that technology had overcome the problems that had undoubtedly existed in the previous century. Marine engines were no longer unreliable, and the problem of instability, which capsized *Captain,* was felt to have been dealt with. These beliefs made the disappearance of the passenger-cargo liner *Waratah* in 1909 all the more puzzling.

Waratah was an unlucky name. A ship of that name was lost in 1848 off Ushant, another in 1887 on her way to Sydney, another in the same month south of Sydney and one in July 1894 in the Gulf of Carpentaria. The latest *Waratah* was built at Whiteinch on the Clyde by the firm of Barclay, Curle and Co. to specifications provided by the owners. She was 465 feet long with a beam of over 59 feet, and a gross registered tonnage of 9339 and a net registered tonnage of 6004—though in the manner of many passenger liners she was advertised as being 16,000 tons because large tonnages attracted passengers. She had five steel boilers to drive her reciprocal quadruple-expansion engines, which gave her a speed of 13 knots, a good speed for those days. Her owners were the Blue Anchor Line, of whom the registered managers were W., F. W. and A. E. Lund. She cost £13,900 and the builders agreed to pay a penalty of £50 a day for each day the delivery was delayed beyond the contracted date. Her specifications were said to be based on those of the steamer *Geelong*, a successful vessel in the same line, and it was hoped she would perform as well as or even better than that ship.

She carried 16 lifeboats for 787 people and another boat which could carry 29, three rafts together able to take 105 people and 930 life belts. Her cabins were luxurious, and she even had a patent distilling apparatus which could provide 5500 gallons of fresh water every day. She carried no radio, but this was not unusual in 1909.

Waratah was completed in October 1908, a splendid ship with a first-class music lounge containing a minstrels' gallery of carved wooden pillars with heavy curtains at each corner supporting post. There were plush settees with tuckered backs, and potted palms, and there was also concealed lighting, *Waratah* being among the first ships to use it.

Captain Josiah S. Ilbery, who was commodore of the Blue Anchor Line and had served it for over 40 years, had been in command of *Geelong* and was sent to supervise the last stages of building. He expected *Waratah* to be his last ship because he was getting on in years and hoped to stay with the ship only as long as necessary before retiring. He was a sound master who had sailed in clipper ships and had great experience of storms at sea.

When she left the Clyde for London, with Ilbery in command, Mr. F. W. Lund, one of the registered managers of the line, was among those on board. When she struck bad weather on the way, it was noticed that she showed an inclination to roll and stick, but the official passenger certificate was issued by the Board of Trade without question. This enabled her to carry 432 people. She was also inspected by Lloyds, who classified her as "+ 100 A1." Since she was designed largely for emigrant traffic between London and Australian ports by way of the Cape of Good Hope, she was also inspected and passed for service by the emigration authorities.

Waratah left England on November 5, 1908, on her maiden

voyage, carrying 67 cabin passengers, 689 emigrants in dormitories in the holds and a crew of 154, including extra stewards who were probably emigrants working their passage. The voyage was trouble-free. Or at least no trouble was reported, since she met no bad weather and she seemed to behave reasonably well. Alert members of the crew, however, noted that Captain Ilbery had found it difficult to load her evenly, because it was difficult to stow cargoes of different weights loaded at different ports. The problems were aggravated by the fact that a cargo-passenger liner uses a great deal of fuel and water during the voyage, thus changing the ship's stability. However, the pilots who had handled her considered *Waratah* a fine ship though "tender." The only real mishap was a small fire in a coal bunker which was soon dealt with.

On her return to England, Captain Ilbery was apparently not asked for, nor apparently did he make on his own initiative, any report on the ship, though, during the course of conversation with his friends, he appeared to be well satisfied. He had formed the opinion that she was not as "stiff" in standing up to the sea as he would have wished and it was known that it was inadvisable to move her in dock without ballast, but this was not considered enough to call for an examination. Either with or without Ilbery's authority, however, the owners told the builders that Ilbery did not consider *Waratah* as stable as her sister ship, *Geelong,* but Barclay, Curle and Co. found no difficulty in satisfying them about her steadiness whether loaded or light.

After being dry-docked and surveyed by Lloyds, and having some minor repairs, *Waratah* left London again on April 27, 1909. The captain, all the officers except one and all the engineers except two had signed on for the second voyage. This time, however, she was far from full with only 22 cabin passengers and 193 emigrants in addition to her crew of 119. Again there was no bad weather and after calling at Australian ports and taking on a full cargo of 6500 tons, mainly farm produce such as wheat and hides and some 1000 tons of lead concentrates, together with 200 passengers, she left Adelaide on July 7, reaching Durban, South Africa, on the 25th. There she coaled and took on a further 248 tons of cargo, making a total cargo load of slightly over 10,000 tons. Her flags flying in the breeze, she put to sea again on the winter's evening of July 26 with 92 passengers on board. The weather report showed a light northeast wind.

One passenger who had traveled from Australia in her, however, Mr. Claude G. Sawyer, an English engineer and businessman, preferred to remain in Durban, though his passage was paid for and he had no affairs in South Africa. He was not new to ocean travel, having made 12 trips in other ships, but during the voyage from Australia, it appeared he had had recurring bad dreams. Night after night they had brought him to terrified wakefulness. He saw himself at the rail of the passenger deck when suddenly from the sea rose the figure of a medieval knight in armor streaked with blood. Blood also soaked a cloth he held in one hand. In the other he held a sword. The mouth opened and shut and Sawyer felt he was saying "The *Waratah*! The *Waratah*!" Then the dream figure sank below the waves. Sawyer was superstitious enough to have believed that the apparition was trying to give him a warning.

The passengers to whom he had told his story had been amused, but Captain Ilbery was not and sent an officer to hint diplomatically that it was not a good idea to spread alarm. Sawyer had no regrets at leaving *Waratah*, however, his only worry being that his wife in England might also think him stupid and alarmist. His telegram to her announcing his change of plan merely stated that he thought *Waratah* was topheavy. Perhaps he had heard others say this, because it was certainly not a new criticism. There was some talk, too, that *Waratah* had rolled in a violent and unusual manner after passing through Port Phillip Heads after leaving Melbourne. But other passengers said they were so pleased with her they were prepared to book a passage back to Australia in her.

Meanwhile, *Waratah* was sailing south with 211 people on board, her next port of call Cape Town, 600 miles away. At about six on the morning of the 27th she was spoken by *Clan MacIntyre*, a smaller and slower steamer, which had left Durban ahead of her for East London. As *Waratah* overhauled and passed her abeam of the Bashee River, she asked the bigger ship her name by lamp.

"*Waratah,* for London."

"*Clan MacIntyre,* for London. What weather did you have from Australia?"

"Strong southwesterly to southerly winds across."

"Thanks. Good-bye. Pleasant passage."

"Thanks. Same to you. Good-bye."

It was a normal sort of conversation between ships passing each other in mid-ocean.

Waratah remained in sight until 9:30 A.M., a matter of three and a half hours, by which time she had left the other ship far behind. As *Clan MacIntyre* watched her hull slip below the horizon and the last thin trail of smoke fade away, she had no idea that she had just had the last real sight of the Blue Anchor liner.

The smaller ship plodded on, like *Waratah* contacting the lighthouse at Cape Hermes for the weather. The lighthouse stands, unnaturally white, against a dark cliff with a few thatched seaside cottages which did not exist in 1909. When called up by *Waratah* and *Clan MacIntyre,* the keepers reported that the weather was hazy but fine, and gave the barometric pressure and the wind. Cape St. Francis also reported a gentle northeast wind. On that coast, however, during a northeasterly wind, it is possible for the pressure to drop and the wind to die abruptly. Then up goes the pressure, the wind shifts like lightning to the southwest and becomes a full gale at once. This is what happened on this day. With *Waratah* about 12 miles offshore, she ran into a head sea. *Clan MacIntyre* met a heavy southwesterly gale which lasted until evening.

That same day at 5:30 P.M. a small ship, *Harlow,* steaming northeast along the east coast of South Africa one and a half to two miles from the shore, saw on the horizon the smoke of a steamer. There was so much of it that her master, Captain John Bruce, remarked to the mate, Robert Ovens, "Damned if I don't think she's on fire." When darkness fell he saw, in the direction where he had seen the smoke, two masthead lights and a red light astern. Two hours later, when *Harlow* was a mile off Cape Hermes, the lights appeared to be 10 or 12 miles away and seemed to belong to a fast steamer coming up behind. Bruce went into the charthouse to check the chart, and when he returned he saw two bright flashes astern, one about 1000 feet high and one about 300 feet high. He thought they were caused by an explosion, but the mate, who also saw them, considered they were merely brush fires ashore, which were common at that time of the year. The lights of the steamer they had seen a little earlier had vanished. Bruce thought no more about it.

A third ship, *Guelph,* a Union Castle liner, also claimed to have been the last to see *Waratah*. On that same day just before 9:30 P.M., or more than three hours after Captain Bruce saw his lights, *Guelph* was abreast of Hood Point, near East London, and eight miles out, when she sighted a large passenger ship some five miles away. The two vessels began to exchange Morse signals. *Guelph* flashed her name and the other ship replied, but Third Officer Blanchard, of *Guelph,* could not make out the message. The distant lamp was very faint and all he could distinguish with any certainty were the last three letters, which were "T-A-H." Since he had failed to identify the ship, neither he nor the

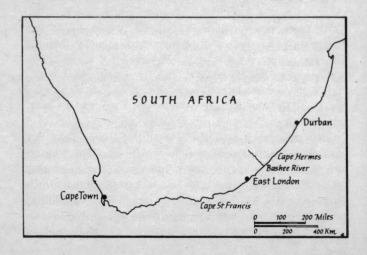

master, Captain James Culverwell, recorded the incident in the log, but, on arriving in Durban and hearing that *Waratah* had just left, he remarked to the fourth officer that she must have been the ship whose signals he had tried to read.

The gale which had troubled *Clan MacIntyre* blew itself out, the wind veering to northeast, but then the next day it blew a hurricane, and for several hours a storm raged which was described as the most severe in those waters within living memory. The captain of *Clan MacIntyre,* who was well aware that Cape of Storms was another name for the Cape of Good Hope, claimed that he had not met anything so bad on that coast during his 13 years as a seaman. The wind, he said, seemed to tear the water up and was of quite exceptional fierceness and power, with tremendous seas. At times, he said, *Clan MacIntyre* had been driven astern, though she was not damaged.

Waratah became due at Cape Town on July 29—and then overdue. The steamship *Illovo* reached Cape Town 24 hours late and the Norwegian steamer *Solveig* arrived damaged on July 31. They had seen nothing of *Waratah,* and anxiety started when other ships arrived which had also left Durban behind her and followed a similar route. Officials talked of "mechanical breakdowns leading to delays," but, strangely, not one of the ships leaving behind her could give any news. The anxiety increased, especially among the companies involved in her insurance. One underwriter told the press, "Were the *Waratah* a less fine steamer, not the same concern would be felt for her safety, but for a twin-screw liner of over 9000 tons, built by a first-rate firm last year, and sailing under good ownership, to be several days overdue is, of course, an unusual event."

Waratah seemed to have disappeared somewhere south of the Bashee River—as completely as if she had never existed. *Clan MacIntyre* reported sighting fully 10 vessels on the 27th and 28th, so it might have been expected that one of them would have reported her. It might also have been expected that, if there had been a disaster, there would have been signs of it in the sea in the shape of spars, life belts, deck chairs or bodies. *Clan MacIntyre,* coming up behind her, and *Guelph,* coming toward her, should have seen something. But there was nothing. Or at least nothing conclusive.

It began to be clear now that some disaster had occurred, but, apart from the reported sightings, there was no notion where it had occurred. Only *Clan MacIntyre* could say with certainty that she had seen *Waratah*. If the ship's lights seen by *Harlow* had been *Waratah*'s, she must for some reason have turned around and been making a reciprocal course back to Durban. What Bruce saw must have been the lights of some other ship and the mate must have been right when he said the flares they had seen were bush fires. The evidence of *Guelph* seemed more convincing, but if the vessel she had spoken to had been *Waratah*, then that ship could have traveled only about 70 miles, as she had passed *Clan MacIntyre* at six o'clock in the morning. Since *Waratah* was a 13-knot ship, this could have been possible only if she had suffered some sort of breakdown, in which case it was odd that she had not been overtaken or at least sighted by *Clan MacIntyre*, coming up behind her.

It seemed incredible that a ship of nearly 17,000 tons, loaded with cargo and carrying over 200 souls, could disappear without trace. Public opinion was so reluctant to accept it that the seas were searched for weeks. The tug *T. E. Fuller* set out on July 31, and on August 1 the tug *Harry Escombe*. Finally, the Admiralty sent out the cruisers H.M.S. *Pandora* and H.M.S. *Forte*, fully expecting them to meet the limping *Waratah*. On August 8, the steamship *Runce* arrived in Cape Town, having changed course to look for *Waratah* but without seeing a stick of wreckage.

Anxiety continued to grow, particularly in Australia where many of the passengers had come from. In Melbourne the churches held special services and ships were asked to keep a special watch for *Waratah*. The Admiralty now brought in the cruiser *Hermes*, and all sorts of theories were offered on *Waratah*'s drift in the event of her having broken down. The ships searched for days, striking far to the southeast on the assumption that *Waratah* had broken down and been carried toward Antarctica by the great Agulhas Current.

Pandora returned on August 10, having sighted nothing. She had experienced extremely heavy weather, but her search had extended 300 miles south of East London. Her commander clearly felt that *Waratah* had sunk and said that if she had broken down east of Cape St. Blaize, she must have been sighted by either *Pandora* or *Forte*. The weather—what was becoming known as *Waratah* weather—consisted of brutal tossing by short quick blows and 40-foot waves with a purple-black sky full of flying clouds. It had driven the search tugs back and hammered *Hermes* for nine days until her hull was strained and she had to be dry-docked.

While everyone was reading of *Pandora*'s search, on August 11 the third officer of the steamship *Tottenham* said he had seen what appeared to be human bodies floating in the sea between the mouth of the Bashee River and East London. One of the apprentices claimed he had seen the body of a little girl in a red dressing gown. The second mate was more exact. The little girl, he said, was wearing a red hood and cape and black stockings. He even added that she was 10 or 12 years old and that her knees were bare. He also claimed to have seen a ship's bunk float by.

On hearing the statements, the ship's commanding officer, Captain Charles Edward Cox, had put the ship about at once and returned to where the traces had been reported. He did not lower a boat, he said, because he was satisfied that the alleged bodies were nothing but sunfish or skate. He personally had seen nothing resembling human remains. His chief engineer agreed and said that "the little girl in the red cloak" was in fact a large roll of printing paper with a red wrapping around the middle. "As we were putting back," Cox went on, "I saw several pieces of what looked like blubber, but nothing in any way resembling a human body," and this made sense because whalers often dumped unwanted offal in the sea off the South African coast.

Despite this, however, on the same day and apparently in roughly the same place, the captain, one of the officers and the stewards of the steamship *Insizwa* also saw what they thought were bodies in the sea, and it was decided, therefore, to send a government tug from East London to the mouth of the Bashee River. None of the reported bodies was seen, only whale blubber and offal.

When *Insizwa* reached Cape Town, a Cape Town newspaper-man called Blenkin was the first to board her, in the dark of the early morning as she lay out in Table Bay. Interviewing the master, Captain W. O. Moore, as he was dressing in his cabin, he was the first to hear the story.

> There were birds flying as far as the eye could see [Moore said], flying in the same direction. That happened at nine o'clock in the morning. . . . I have not the slightest doubt that they were the bodies of human beings. Two were clad in white, two in dark cloth. There was no wreckage about. The bodies were floating in close proximity to the ship and I was able to see two of them out of the cabin window. The others were observed on inspection from a higher point.

When Blenkin asked the obvious question, why a boat had not been lowered to pick up the bodies, Moore said, first, that he had lady passengers on board and didn't wish to alarm them, and second, that the cargo had shifted, giving the ship a slight list so that, in the heavy seas running, it would have been dangerous to bring her to.

After returning ashore, Blenkin reported the substance of the interview both to his newspaper and to Sir David Graaf, Colonial Secretary. At first this new evidence was accepted without question, but when he saw his statements in print, Moore began to back down. Blenkin immediately suspected that there was more behind the captain's retreat than mere doubt.

> On board the *Insizwa* [he wrote] was someone in authority in the company that owned the vessel who immediately realized that if Captain Moore persisted in his statement both he (the captain) and the steamer would be held up and the service of the line disorganized by the compulsory attendance of himself and some of the crew at the Cape enquiry which had been forecasted.

It was sound reasoning because such a thing could have cost the company a lot of money. Captain Moore became more and more vague as the days passed and not all his officers agreed with him anyway, though the belief in the bodies was strengthened when the steamer *Bannockburn* arrived on August 19. She had run into weather similar to that which had struck *Clan MacIntyre* and her captain was in no doubt about what had happened to *Waratah*.

The day that *Pandora* had returned, however, the agents of the Blue Anchor Line had received a telegram from East London saying that a Blue Anchor ship had been sighted at a considerable distance from the land making slowly toward Durban. Immediately the word went out that it was *Waratah,* but the steamer passed Durban the following morning, 10 miles out and, though the weather was hazy and the ship was not recognized, she was clearly not *Waratah*. Hopes were again raised when *Port Caroline* was sighted four days later. It was firmly expected she would have *Waratah* under tow, but she came in alone, having seen nothing of the missing ship.

The next story to appear stated that *Waratah* had left Durban with coal on her decks. This was cheaper in some ports than others and it was not unknown for ships to buy it where they could and transport it on their decks. Perhaps heavy weather had shifted it. But the captain of *Clan MacIntyre* had seen no coal on *Waratah*'s deck and others agreed.

The mystery deepened. *Waratah* had provisions for a year, and the Australian government, believing her to be drifting, paid for the ship *Severn* to hunt for more than a month and sail 2700 miles. The most thorough search of all was made by the Union Castle steamer *Sabine,* chartered by the Blue Anchor Line. On board she had an officer and 75 naval ratings, searchlights, towing apparatus and extra stores. She left Cape Town on September 11, returning on December 7, having cruised for 88 days and covered 14,000 miles through the seas and islands of Antarctica. In 1899 a ship named *Waikato* had been disabled in the same waters where *Waratah* had last been seen and had drifted southward for 14 weeks, eventually being found near the lonely island of St. Paul, more than halfway to Australia. *Sabine* followed the route that *Waikato* had drifted, but without result, save that at St. Paul she found relics of other missing ships. With not a stick of wreckage found, the Admiralty called off their search and on December 15, 1909, *Waratah* was finally posted missing at Lloyds. When 1909 became 1910 no one believed she could still be afloat, but hope was not finally abandoned and in February, following a public meeting at Melbourne, the steamer *Wakefield* was sent out on a systematic search lasting three months.

With no news and no hope of news, the inevitable clairvoyants got in on the act. One, a woman called Morris living in King William's Town, South Africa, who was reported to have known nothing of *Waratah* at the time of her disappearance, was said to have had a vision of her, three miles off her course in a storm, striking an uncharted rock, heeling over and plunging down, funnel first, in a matter of seconds. The world's newspapers also produced their share of wild stories and speculation. One claimed it had evidence that white children had been washed ashore and were being cared for by African coastal tribes. Certainly there had been children on board, five in one family, who, despite Mr. Sawyer's fears, had boarded her at Durban. There was not a word of truth in the stories, however—though by this time Mr. Sawyer was enjoying a certain notoriety.

Rumors continued to arrive. People who had said nothing until the ship was finally written off now came forward claiming they had always known she was unseaworthy, that she carried too much top-hamper, that the captain or one of the officers had said she was the most unstable ship he had ever sailed in and that she had nearly capsized even in dock. One crew member, it was claimed, had said he wouldn't sail in her again for 10 times his pay. She had been engulfed by a freak wave or by an earthquake; pirates had boarded her and killed everybody—all 211 of them; she had run aground in Adelaide; her decks had almost broken loose during her maiden voyage; her boats were rotten; and she pitched and rolled abnormally, even in calm seas. In some of the stories there appeared to be elements of truth, but no more. No fewer than five bottle messages were found on Australian beaches within a year, all supposed to have been from *Waratah*, and a piece of timber clearly marked "S.S. *Waratah*" was found on the shore near Durban. Every one was found to be a hoax.

One story did gain strength: that *Waratah*'s stability was doubtful, and people were even prepared to swear she had a heavy list when she left Port Natal.

The Board of Trade inquiry was opened at Caxton Hall, London, on December 16, 1910, over a year later, even as a false report came in from East London that deck chairs marked "*Waratah*" had been washed up at Coffee Cove on the South African east coast. It was also stated that a bottle had been washed up near Brisbane containing a message, "At sea, latitude 40 S. *Waratah* broken down. Drifting south. All well but anxious." It was signed "J. G. Jones" and asked the finder to contact Mrs. J. G. Jones, of 58 George Street, Sydney. Unfortunately a search revealed there was no such person as Mrs. Jones.

A model of *Waratah* had been provided, with a removable top that showed her music lounge and minstrels' gallery, but in the absence of a single survivor, collecting evidence had been difficult and many of the witnesses were seamen who had had to come from Australia or South Africa. The president was Mr. J. Dickinson, a stipendiary magistrate, who was assisted by four assessors, among whom were an admiral and a professor who was also a naval architect.

Immediately it became clear there would be gaps in the evidence. First of all there was no plan to show how the cargo had been stowed when *Waratah* left Durban. One had been prepared, based on the manifests of agents, stevedores and Captain Ilbery's reports about the cargo shipped in Australia, but it had been in the possession of Chief Officer Owen and had gone down with him when the ship had vanished. There was also no real evidence on the ship's stability and none at all on the last voyage. What evidence there was proved to be conflicting and some of it even seemed evasive. It soon became clear that while the court was chiefly concerned with the matter of stability, the owners were equally concerned that whatever else might be blamed for the loss, this should not. It was a normal enough attitude to take, of course, because their reputation was at stake.

The expert witnesses were almost unanimous. They claimed that the ship was properly designed and constructed, that she sailed in a seaworthy condition, that she was adequately manned and that her rigging and lifesaving equipment were in good order. They were especially positive that her loss could not be attributed to faults in her construction and pointed out that she had passed five separate inspections—the builders', the owners', the Board of Trade's, Lloyds' and the emigration authorities'. Three of these inspections had been by independent authorities who had no interest in overlooking weaknesses, four were of a technical character and none had reported the smallest defect. Her owners, the builders reported, were so pleased after her first voyage that they had asked for a copy of her plans to be considered for further use in the building of new ships.

Thomas Miller, Assistant Chief Draftsman to Barclay, Curle and Co., even said that if the vessel lay over she would still right herself, and that 290 tons of coal placed on her side would make her list no more than 15 degrees. Complaints they had received about her did not concern her behavior at sea but her tendency to shift in dock without ballast. The firm had built almost 500 ships without complaint.

The expert evidence was supported by Sir William White, formerly Chief Constructor of the Royal Navy, and Mr. Robert Steele, a naval architect. White declared that the stability figures for *Waratah* showed that she was not topheavy and he was confident that the "list" and the abnormally long roll on which several people had remarked could have had nothing to do with the disaster. Steele was convinced that, unless something had happened which had caused the sea to enter her holds, no violence of the wind or sea could have capsized her.

Nevertheless, while the experts were prepared to show what had not happened, none of them could produce any explanation of what had. Some of them even seemed to feel that sinking was impossible. Yet if *Waratah* was a perfectly stable ship, why should she have disappeared during a storm which was weathered by plenty of other ships in the same waters?

A statement by Mr. F. W. Lund about her behavior on her voyage from the builders' yard to London was challenged. Lund claimed she had behaved splendidly. "She rolled a little when she had had to wait off Dungeness for a full hour," he admitted, but she had only a slight list when broadside to a gale. Third Officer Henry had apparently not agreed, however. He had been lost in *Waratah* but he had told a Mr. Latimer, of Sydney, who now reported what Henry had said, that she had frightened him at Dungeness because he had thought she would roll over.

A great many amateur witnesses with little or no technical knowledge were called. Many of their statements were proved to be contradictory or even wrong. Some were merely nervous passengers not used to the sea, and in the 18 months since *Waratah* had disappeared memories had become treacherous. Among them, however, was one who might be considered an expert, and he had a great deal to say on the matter. He was Professor William Bragg, a Fellow of the Royal Society who held the chair of physics at Leeds University and was later a Nobel Prize winner for physics. He had joined the ship in Australia on her first voyage homeward and had been alarmed by her behavior. One morning, he said, the starboard list was so pronounced that water would not run out of the baths. The list continued for several days and he didn't think it had anything to do with the wind. A water ballast tank, he said, was filled to correct the list, but instead it merely made her list to port.

In his opinion, the vessel was in "neutral equilibrium" and, knowing something of the subject, he had asked the captain for her stability curves. But Ilbery had said he had not been furnished with the information and there were no such data aboard.

"My impression," Bragg continued, "was that the metacenter was just slightly below the ship's center of gravity when she was upright, then as she heeled over to either side she came to a position of equilibrium, where she hung for a considerable time." It seemed to him that *Waratah* had not a sufficiently correct balance to bring her upright when she rolled and it was a matter of pure luck that she hadn't been lost on her maiden voyage.

Despite this, however, and despite claiming that he had been told by one of the junior engineers that she was "the tenderest ship he had ever been in," he was sufficiently reassured after a talk with the chief engineer to continue home in her.

Bragg's evidence was strong, but the owners produced more experts who were prepared to take an opposite view, and plenty of passengers who swore on oath that they had noticed nothing wrong. David Tweedie, who claimed to have made 16 ocean voyages of one sort or another, said he had never been in a better ship or had a better voyage. He had, he said, heard no complaints from Professor Bragg, near whom he sat at meals, and all the usual deck games were played. But the value of his evidence was diminished by the fact that he was described in court as a "personal friend of the owners," while on the other hand, Coal-trimmer William Marshall had not liked the way *Waratah* rolled and he had gotten out of her in a hurry.

Able Seaman Nicholas said that when he was signing the ship's articles, Chief Officer Owen had said to him, "If you can get anything else, take it, because this ship will be a coffin for somebody." He too had made a point of leaving the ship in Sydney.

Able Seaman Edward Dischler, a man with a long experience of deep-sea ships, including the North Atlantic and the Australian runs, said that when *Waratah* rolled she went right over and didn't seem able to recover herself but stayed there quite an appreciable time, and the portholes had had to be kept closed to prevent water coming in as she rolled. "She was the unsteadiest ship I ever voyaged in," he said. The ship's surgeon on the first voyage, Dr. Harold Skarratt Thomas, said much the same thing, complaining of a heavy list and a sudden jerk which had upset him one morning in his bath. He also had left *Waratah*.

Steward Frederick Little, who had left the ship in Durban to obtain employment there, stated that she had rolled heavily, that there was a lot of talk about her among the officers and stewards and that he had heard some of the stewards express the opinion that she was topheavy. He, however, had seen nothing abnormal. Another steward by the name of Herbert, who had been on the maiden voyage, claimed that *Waratah* had a pronounced list. "She rolled excessively, much more than *Geelong*," he said. A great deal of crockery had been broken and, worse, the promenade deck moved about on its beams when the ship rolled. He claimed, "The bolts supposed to hold the deck planks down were broken," and fell on the baker's head. It was possible to put a finger in between the deck planks and one of the beams to which they should have been bolted. Samuel Lyons, another steward from the first voyage, had been told by the boatswain, "By God, I wouldn't like to be in this ship in a storm. She'd go to the bottom," and a passenger named Duncan Mason, who held a first engineer's certificate and had been 33 years at sea, had been so alarmed by the unsteadiness of the ship that he had spoken to the chief officer, who was a friend of his. "Owen," he said, "if I were you, I would get out of this ship. She will be making a big hole in the water one of these days." To which the mate replied, "I am afraid she will."

There were, however, many passengers and former members of the crew who took a totally opposite view. These said they had not noticed a list or that it was so slight as to be negligible, that the rolling was never excessive and that *Waratah* was as comfortable and seaworthy as any ship they had sailed in.

Harry McKay Bennett, a former deck officer, claimed that her behavior on her first voyage had shown an easy roll and no abnormal list. "She neither pitched nor rolled anything out of the ordinary," said John F. Ryan, who had been fourth engineer on the first voyage and had left on promotion to be third engineer in another ship.

A passenger on her maiden voyage, Mr. Worthington Church, held a different view. "I have made three voyages to Australia round the Cape," he said. "I thought she was very topheavy. I had a conversation with Captain Ilbery, who said he was not altogether satisfied with the ship." Another passenger, Leslie Wade, said the vessel rolled so much that the piano in the passengers' lounge moved, and her movement was so unpredictable that on the way from the Cape to Sydney the fiddles (the small wooden containers to stop crockery sliding off) were on the tables the whole time. Mr. Morley Johnson, also a passenger on the maiden voyage, however, said, "The behavior of the vessel was in general quite equal to that of any vessel I have been in. Her rolling was not unusual. No person expressed to me any doubt of the ship's stability."

Other passengers gave similar evidence, and the fiddles could well be explained by the fact they would always be used in that stretch of the ocean between the Cape and Australia where big seas run whatever the weather and any ship, whatever its condition, could be expected to roll heavily.

The truth seemed to be that *Waratah* had not behaved too badly, though she had not experienced any really bad weather, but the impression remained that there *was* a list and that she had rolled sufficiently to attract the attention of many people.

It was suggested, in an attempt to explain the sudden disappearance, that perhaps the steering gear had broken down during the storm and that the ship had been overwhelmed by the seas before the hand-steering gear could be connected. But Mr. James Shanks, who had been superintendent engineer to the owners for 13 years, said that the change would have taken only four to five minutes, while, since *Waratah* was twin-screwed, she could have been steered for a time by means of her engines.

Then came the turn of Mr. Claude Sawyer. Despite his dreams, he seemed a very stable person, and he was one of the few survivors from *Waratah*'s second voyage from Australia. He had no knowledge of the technical or seafaring terms he had heard, but he had noticed the list even as she had left Melbourne. When she rolled, she recovered very slowly, then came up with a jerk, and several passengers had been flung about and hurt. He had heard the third officer, he said, express the opinion that the ship was topheavy and that he intended to leave her if it didn't mean he would have to leave the company. One morning, Sawyer went on, he judged from the level of the water in his bath that the ship was lying over at an angle of about 45 degrees, and he was so concerned he spoke to one of the officers, who attempted to reassure him.

Curiously, however, it seemed to have been the pitching which worried him most and he had spoken about it to a fellow passenger, a solicitor called Ebsworth who had once been a ship's officer himself. Ebsworth, he said, was equally concerned about the ship and they had walked together along the promenade deck and watched the movement of the ship as she rose and sank in the swell. When she was in the trough between waves, instead of rising to the next wave, she plowed through it, taking on a lot of water, and seemed very slow in recovering.

Sawyer had been so concerned that he began to think of leaving the ship at Durban, but the voyage was uneventful and he had almost forgotten his intention until three or four days from Durban when the alarming dreams started.

"I saw a man in a very peculiar dress . . . with a long sword in his right hand. . . . In his other hand he had a rag covered with blood. I saw that three times in rapid succession during the same morning, and the last time . . . I could almost see the design of the sword." It had been enough for Sawyer, and he had decided to quit the ship and had even tried to persuade Ebsworth to leave with him.

Even while in his Durban hotel the dreams did not stop, he said. On July 28, the night *Waratah* disappeared but before she was reported missing, he had had another dream. "I saw the *Waratah* in big waves. One . . . went over her bows and pressed her down. She rolled over on her starboard side and disappeared."

Sawyer never varied his story nor tried to elaborate on it and it had a curiously convincing ring, because he had told it to the manager of the Union Castle Line in Durban, with whom he had booked a passage home even before *Waratah* vanished. Quite obviously his fears about the ship had so preyed on his mind as to produce the nightmares, and even skeptical seamen felt that there must have been something odd about *Waratah* to so affect a man who was an experienced traveler and by no means hysterical.

The court was impressed enough with Sawyer to praise him for his integrity and courage in coming forward, and inevitably the name given to the character he had seen was Vanderdecken,

the Flying Dutchman, the blasphemous captain condemned to
sail the Cape seas through eternity. Yet his evidence about Ebsworth's being concerned does not appear to be entirely convincing because Ebsworth had not been persuaded to leave the ship
and had been lost with her. Indeed his widow produced a letter
he had written from Durban saying that *Waratah* was "a fine
sea boat."

Nevertheless, Sawyer's story about the pitching was supported
by Mr. George S. Richardson, Chief Mechanical Engineer of the
Geelong Harbour Trust, though his observations on it had originally been in the nature of a joke. He had said to Sawyer one
day after some particularly bad rolls which had been accompanied by pitching and trembling, "One of these days she will dip
her nose down too far and not come up again." In fact, he didn't
believe what he said and would not even have remembered it if
the ship had not disappeared. Yet, between them, Sawyer and
Richardson had produced a new angle. In a heavy sea any ship
will pitch, tremble and shudder, sometimes damage can be
caused, but it seemed unlikely that the ship could have continued
her downward sweep enough to disappear.

The court was thorough and listened patiently to everyone who
had anything to say. Even William Saunders, a stowaway on the
first voyage, who had been scared by the rolling, was heard. Yet,
for everyone who said there was something wrong with the ship,
there was another who denied it. While one passenger said the list
was so great the people on one side used to shout to those on the
other, "How are you up there?," another claimed that he saw
nothing wrong. One coal-trimmer on her first voyage thought
the ship had rolled a lot; another said everything was fine. One
seaman said that the ship listed; another said she sailed quite
normally.

The court was told of the master of the P. & O. steamer *Mon-
golia,* who was reported to have been informed by Ilbery that this
would be his last voyage in *Waratah* unless she were materially
altered when she returned to England. But the captain of *Mon-
golia* later telegraphed the court to say that he had never met
Ilbery in his life. The court was even prepared to hear the master

of a ship called *Talis,* who wrote claiming he had seen *Waratah* and knew what had happened to her, but this correspondent disappeared before he could be brought before the court. It was subsequently discovered there was no such ship as *Talis.*

Steward B. J. Shore said that the dissatisfaction with the ship was chiefly among the old sailors, and the impression is gained that a lot of them were exaggerating the dangers to newcomers and to passengers in the way of all old sailors, to indicate their own experience, and that there were many aboard who would not have given the rolling or the list another thought if the ship had not been lost. Yet there were experienced ocean travelers who weren't happy. One of them, Mr. A. W. Sedgwick, who thought *Waratah* seemed very high out of the water, had spoken to Chief Officer Owen who had asked him, "Do you think she's topheavy?" His reply had been, "I don't think it. I know it. I'm not a landlubber." Another officer to whom he spoke agreed with him and told him the reason was that the decks were 10 inches to a foot higher than in other ships of her size. He had felt sure she would turn turtle.

At one point the court was sidetracked by Nicholas Sharp, a seaman on the first voyage, who said the ship was never straight and hard to steer, and that the boats were faulty and were covered with thick paint which was jammed into the seams. In this his testimony was confirmed by Alfred Pinel, the carpenter's mate, an ex–Royal Navy man, who said the boats were built of green wood and that the davits couldn't be shifted. But he insisted there was nothing wrong with the ship herself, though he did agree she had a list and that some of the seamen on her first voyage described her as being "dead on the roll," which was much the same thing as Professor Bragg had described in more technical terms.

Even the evidence of the tally clerks in Sydney did not agree. One of them, John Latimer, said he was told by one of the officers that *Waratah* had one deck too many; yet another, Alan George Melville, said that Chief Officer Owen had told him he had never been in a better ship in a seaway. "She's like a rocking horse," he had said.

The men who had last seen *Waratah* appeared. The pilot who had seen her passing through Port Phillip Heads after leaving Melbourne thought she rolled violently and in an unusual manner; but the master of the tug *Richard King,* which had pulled her off the wharf in Durban, reported that she had seemed to be perfectly all right to him and did not appear to act like a tender ship. The master of *Clan MacIntyre,* who had been the last man to see her, reported that she appeared neither to have a list nor to be rolling excessively, but was proceeding in an exceedingly steady manner. In this he was supported by his first officer, but one of his apprentices, Stephen Lamont, said that when he saw *Waratah* she was sailing like a yacht, heeled well over and pitching enough to show her propellers.

The master of *Guelph* described how they had picked up the message of which only the letters "T-A-H" had been decipherable. But, he admitted, he was less than 100 miles from Durban at the time and *Waratah* ought to have been much farther south. The master of *Harlow,* who had seen the flashes off Cape Hermes, said he had felt at first that *Waratah*'s bunker coal had overheated and she had blown up. But there had been no sound of an explosion, and the lighthouse keeper ashore had seen no flashes or fire and had heard nothing, while *Harlow*'s chief officer had seen no rockets, which he felt he would have done if *Waratah* had been on fire. They had attached so little importance to the incident, in fact, that neither had made a report of it or recorded it in the ship's log, thinking no more of it until after learning that *Waratah* was missing, when one of *Harlow*'s officers, Alfred Harris, thought the lights he had seen might have been hers.

Then the men from *Tottenham* gave their evidence and once more there seemed to be considerable disagreement. Apprentice M. W. Curtis reported seeing objects streaked with red in the water. He later heard these were sunfish, but he still believed one of them was the body of a little girl wearing a red dressing gown. Mr. T. Stewart, the second engineer, also thought he saw the body of the girl, and soon afterwards, just below the water, what he thought was the trunk of another body, and then a floating mattress and sheet, so that a Chinese fireman observed, "Plenty dead bodies sea side." Engineer Artificer A. J. Tucker also saw what he thought was the body of the girl. Third Officer E. F. Humphrey also saw the sheet and what he thought were bodies, but Chief Officer David Evans thought that what he saw was blubber, and Chief Engineer John Hammond considered that what *he* saw was a roll of paper with a red wrapper.

What did become clear, however, was that they had all known *Waratah* was missing and had expected to find bodies, and that Captain Cox, like Captain Moore, of *Insizwa,* had not intended having his ship delayed by having to give evidence. Though he personally had thought that what had been seen were sunfish, he had also told his crew that they were not to report anything.

When the men from *Insizwa* gave their evidence, it began to appear that while *Insizwa* was off the Bashee River *Tottenham* was almost 100 miles away, south of East London, so that they could not have seen the same things, and that, anyway, since *Waratah* had last been seen off the Bashee by *Clan MacIntyre,* it was surely to be expected that any bodies found would be farther south. In fact, the court did not feel that what *Insizwa* and *Tottenham* had seen were bodies.

One of the most significant points that emerged was that no report on the ship's behavior from Captain Ilbery was offered, and it even began to appear that all mention of the subject was being suppressed. Letters from Captain Ilbery were produced, but in none of them did he mention the ship's behavior, and the court remarked acidly that it seemed very odd, because although the letters mentioned every other sort of triviality, at no time did Ilbery, a highly experienced master, draw the owners' attention to the one thing they would be expected to wish to know about, especially as *Waratah* was for them a new type of ship and they claimed to be contemplating using her specifications as the basis for another.

"The court is quite unable to understand how silence could have been preserved on such an important and interesting subject as the *Waratah*'s stability and behavior at sea," the president commented. It was contrary to the whole practice of the sea. The court was clearly implying that they suspected the owners of holding something back, even that they had deliberately suppressed Ilbery's report because it reflected badly on them.

The belief was strengthened by correspondence which had passed between the owners and the builders. After *Waratah* was loaded with a full cargo for the first time, the owners had sought a meeting with the builders for which it was clear there must have been a good reason. In a letter to the builders, the Lunds had commented, "From what our representatives report, it seems clear that the *Waratah* has not the same stability as the *Geelong*," and in another letter, dated April 4, 1909, they had said, "We have consulted Captain Ilbery and he has been able to convince us that this vessel has not the same stability as the *Geelong* and, considering that he was present during the construction of both vessels and has commanded them both, he is in a perfect position to judge this and all other matters." But the owners could produce no evidence that Ilbery had said anything on the subject of stability or the reason for the conference. And what were the "other matters" the second letter mentioned?

Questioned, they insisted that it was *Waratah*'s stability in port when working cargo and bunkering that had worried them. The court was unconvinced but, though it expressed quite firmly the view that it was not being told the full story, it got no further information, and the owners insisted that the question of stability arose only because they were fighting for their demurrage—the agreed £50-a-day penalty, because the ship had not been delivered by the contracted date. Their complaints about stability, they insisted, were pure bluff. The court replied with a demand to know when the demurrage was paid, but the owners could not say.

Observing that, when so little was known and the range of conjecture was so wide, it was idle to discuss the many guesses put forward by clairvoyants and amateur theorists, the court retired. The two most important points seemed to be the evidence of Professor Bragg and the fact that the owners had produced nothing to indicate what Captain Ilbery thought of the ship.

The court delivered its findings on February 22, 1911. *Waratah* had last been seen by *Clan MacIntyre,* heading in a direction which would take her into a position where she would feel the full force of a storm, but they could not feel she had been merely disabled because otherwise she would have been seen again by the *Clan* ship or picked up by one of the ships looking for her later. She had proper and sufficient boats and lifesaving appliances, all in good order and ready for use; the cargo was properly stowed; and the ship was in proper trim for the voyage and in good and seaworthy condition with regard to structure; but there was not sufficient evidence to show that all proper precautions as to battening hatches, securing ports, coaling doors, etc., had been taken, and it could only decide that the ship was lost suddenly in a gale of exceptional violence, which had caused her to capsize.

The court did not commit itself on the reason for the capsizing, however, though it did comment unfavorably on the fact that Captain Ilbery had apparently not been asked to report on *Waratah*'s stability or behavior at sea after her maiden voyage. The court was compelled to draw an inference unfavorable to the owners from Ilbery's statement that *Waratah* was not as stable as *Geelong* and from the correspondence between the owners and the builders on this point. It also recommended that there should be a further study of the question of stability in ocean-going ships, a recommendation repeated 30 years later in the case of the lost tramp *Anglo-Australian* and 10 years later still in the case of the lost *Hopestar*.

Blue Anchor Line, though not condemned, were clearly criticized, and they decided to sell their other ships, *Commonwealth, Geelong, Narrung, Wakool* and *Wilcannia,* to the P. & O. Company, which used them, oddly enough, to inaugurate their "branch" service to Australia.

What had happened to *Waratah*? It was considered she could not have struck a reef, because there would have been wreckage, and she could not have blown up or caught fire because she would surely have been able to fire flares and get away boats. Though *Waratah* had long gone, the questions remained. If she had sunk, bottom upward, pinning loose or insecure gear under the main hull, she would surely have settled on her side, thus eventually releasing the loose gear. Yet nothing was ever found, even four years afterwards, and stories continued to crop up about what had happened.

Inevitably there were reports that *Waratah* had carried treasure. In fact there was none. She carried a cargo of frozen meat from Australia, some ore, 2000 tons of bunker coal and 279 tons of fresh water. In 1922 a man called John Noble announced that, although not aboard *Waratah* when she left Durban, he transferred to her in mid-ocean from *Telemachus* to act as chief stoker. When *Waratah* capsized, he managed to reach the shore and went immediately to Cape Town, but his story of the disaster was not believed. He then took ship in *Themistocles* and spoke no more of the incident until 1922. He was soon proved a liar because *Themistocles* was not built until two years after he said he had joined her, and, when *Waratah* disappeared, she was several thousands of miles from where *Telemachus* must have been.

In the 1920s a man who had spent most of his life in South Africa told J. G. Lockhart that shortly after the ship's disappearance a man was found wandering in the veldt close to the coast and not far from Port Elizabeth. He had nothing by which he could be identified, and could give no account of himself other than a vague intimation that he had come ashore from *Waratah*. No further information was obtained from him and, since he was clearly out of his mind, he was sent to the Grahamstown Asylum. Another story Lockhart obtained came from a man named Carter who claimed he had had a hand in *Waratah*'s loading and that on the afternoon she sailed, Ilbery had shaken hands with him and said, "I do not want to go, but I will keep my promise and take her." It was his view that *Waratah* was a tender ship, very high out of the water, and that coal was dumped on her deck. Running into the storm before the coal could be stowed had caused her to turn turtle. It was true *Waratah* had a coal bunker on her spar deck, but even had the bunker been full it would not have affected her stability to any serious extent, and several people had testified that she had no coal on her deck.

Yet another story came from a man called H. D. Barry, who wrote to the *Daily Mail* saying that while on a visit to South

Africa in 1913 he was in the Transkei, a native reserve between East London and St. John's. It was common knowledge among the inhabitants of Willowvale, he said, that the loss of *Waratah* was witnessed by at least one white man. Barry claimed to have traveled to this man's camp at the mouth of the Qoka River, where he was told that one very wild night a large steamer had been seen close inshore. The man had returned to his hut for his binoculars, but when he regained the shore the ship had disappeared. He knew nothing of the loss of *Waratah* until some three months later, but, when he compared dates, he was convinced that the ship he had seen was she. He was even in possession of a piece of wood with the letters "W-A-R" carved on it.

The story sounded reasonable because Cape southeasters are notorious. When they blow hard, they pluck the surface from the sea, as the captain of *Clan MacIntyre* had noticed, and carry it along parallel with the waves, and there are terrific tides and surf in this area. With the tremendous depth of water it is quite possible that a ship could turn turtle, and there are large sharks which could well account for any human beings who remained afloat.

But a few days later the story was derided by a Mr. Harry Hulse, formerly of the Cape Mounted Rifles, who had been one of those searching for wreckage along the coast.

"It comes as news to me," he wrote indignantly, "that a white man saw a large steamer close inshore." Not only were the white traders and natives living along the coast interrogated, he said, "but not a scrap of evidence appeared to indicate a ship had been sighted near the shore." He had lived in the area for some years but had never once heard of fresh evidence to explain the mystery. He had also visited the particular locality referred to by Barry but had never heard the story before. But another member of the Cape Mounted Rifles, Edward Conqueror, claimed that while taking part in military exercises at Xora Point, south of Port St. John's, he had seen a ship whose description tallied with that of *Waratah* making heavy weather in a gale and then rolling over and disappearing.

Still the stories came. In 1936 Lockhart received a letter from

Mr. Percy Evans, chief officer of *Rabaul,* in which he described an alarming experience in the waters where *Waratah* had disappeared. He was in a well-found steamer of 11,000 tons, homeward-bound from Sydney with a cargo of grain. After calling at Durban the ship ran into a southerly gale which soon reached hurricane force. Steam pipes and steel ladders were torn away and deckhouses stove in, and the starboard boat was crushed. At 3:00 A.M. a gigantic wave came over the starboard bow, smashing the forward hatches and caving in the saloon and the front of the bridge. For some time the situation was critical. He pointed out that, because of the steep fall in the ocean bed in these waters and because of a warm current meeting cold polar winds, violent but quite local storms occurred. He felt certain that *Waratah* had been struck by a similar huge wave and then a second had come aboard, filling the holds and sinking her.

Some years later an Irishman called Staunton claimed to be a survivor from *Waratah* and for a while enjoyed a certain notoriety. His story was well received until someone thought to look up the records and found there was no Staunton in either the passenger or crew list.

In the winter of 1942 a South African air force pilot, Lieutenant D. J. Roos, spotted a submerged wreck while flying a scheduled mail run over the area where *Waratah* disappeared. When he mentioned it to a naval officer who had conducted a survey of the South African coastline, the naval officer was puzzled because there was no known wreck at the spot where Roos claimed to have seen one. Roos drew a map and described the ship as lying on its side on one of three rocky shelves off Mazeppa Bay. Immediately the naval officer thought it might be *Waratah* and an aircraft was chartered by a Durban newspaper. Accompanied by a reporter, Lindsay Smith, Roos flew over the route, but they were dogged by bad weather and engine trouble and nothing was found. Shortly afterwards Roos was killed in a motor accident and the map he had drawn of his find vanished.

Theories continued to appear. One was that *Waratah* had been sucked into a "blow hole" and been drawn down by the strong inshore counter-current into a vast undersea cavern. Another theory was that a natural vortex had been formed and dragged her down. The Agulhas Current flowing along this coast, and known as the Mozambique Current north of Laurenço Marques, is a warm tropical current which reaches a surface speed of about five knots in this area, with higher underwater speeds. A powerful counter-current could certainly build up to hammer it and create a maelstrom of sea, something which *Clan MacIntyre,* which barely escaped disaster, reported.

In 1954 an Englishman, Frank Price, said that 40 years earlier a Boer named Jan Pretorius had witnessed *Waratah*'s end. He had seen her wallowing inshore with no sign of life aboard her, then she had heeled over and disappeared. He had not come forward before, Price explained, because he had been illegally prospecting for diamonds and to mention having been there would have meant imprisonment for life. He had confided in Price but insisted that nothing should be said until he died. By 1954 Price had been certain that he was dead.

At the end of 1955, Alan Villiers, one of the most distinguished writers about the sea, did a broadcast on *Waratah* for the BBC and had barely finished when a listener rang up to say he had been in Durban when *Waratah* sailed and had seen her round the Bluff. The ship, he said, was heavily laden on her decks with yellow timber which looked topheavy and he had remarked to his wife that he was sure she would roll over. However, the one thing *Waratah* was not carrying was timber.

Nevertheless, Villiers was contacted by surviving friends of Captain Ilbery and Chief Officer Owen. One who wrote from Lewes, in Sussex, said Ilbery was a close friend of his family, and they had sailed to Australia with him in *Wilcannia* in 1901 and home again in *Commonwealth* two years later. After that Ilbery had been a frequent visitor to the house. After the maiden trip of *Waratah,* Ilbery had been less cheerful than usual and one evening had told them he was not happy with the ship. When the head of the Lund family had retired, he said, his sons had taken his place and had built *Waratah* to make larger profits. Ilbery had not liked her.

Another man, a former shipping official, who claimed to know the father of Chief Engineer Hodder, who had died in *Waratah*, said that as soon as the ship became overdue, the old man had been certain she had capsized. In their last conversation, he said, his son had told him, "I'm not going back in her after this trip. She's topheavy."

It was Alan Villiers's opinion that the tragedy was one of stability. The owners, acting in good faith, gave the builders a very difficult specification and, though everybody did his best, the result was not good enough and there was not time to learn how best to load the unstable vessel before she struck her first severe storm. He felt she rolled over in the gale and, with five steel boilers breaking from their beds and the roar of steam stifled by the inrush of the sea, the passengers barely had time to scream.

K. C. Barnaby's view was more technical but reached the same conclusions. It was obvious, he wrote, that *Waratah* was an extremely tender ship with little positive stability. Tender passenger ships were by no means unknown at that time, because designers were reluctant to provide more than a bare minimum of stability. The average passenger could not distinguish between a quick, uncomfortable but safe roll and a long, slow comfortable one that was dangerous, and owners were afraid of their ships' acquiring a bad reputation for violent movement.

A tender ship requires the utmost care in cargo stowage, careful handling and no risk of large quantities of water being trapped on a deck without ample freeing ports. Given these and other needs, he went on, certain Atlantic liners plied the seas for their full period of life without accident. It was much easier, however, to regulate stowage on a passenger liner with little or no cargo than on a ship like *Waratah* with a large cargo capacity.

The known facts of tenderness and the absence of wreckage, Barnaby felt, pointed to the conclusion that *Waratah* was capsized by being caught by a heavy sea as she neared the point of no return. It was also possible that a shift of the heavy lead concentrates in her cargo contributed to or even caused the disaster. When a ship capsized and sank bottom upward, loose or insecure gear became pinned under the main hull and did not float clear. It was this very absence of wreckage, he felt, that pointed to a capsizing. It is now compulsory for all passenger ships to be inclined, and any ship without a reasonable metacentric height is compelled to carry permanent ballast, while the introduction of stabilizers has also contributed to safety as designers are no longer so tempted to secure a gentle, slow rolling by a dangerously low metacentric height.

Other views appeared, one of them that of Captain W. S. Byles, of the 28,000-ton *Edinburgh Castle*. Writing of an experience during the 1960s, he said that while in heavy seas in the area where *Waratah* disappeared he placed the swell on *Edinburgh Castle*'s port bow so that the ship rode comfortably. The distance between the waves at the time was about 150 feet, but suddenly the next wave appeared 300 feet away—twice as far—so that when the ship pitched she charged head-on into it at an angle of more than 30 degrees, shoveling it aboard as *Rabaul* had done, before she could recover, almost as if she had dropped into "a hole in the ocean." A retired seaman, Mr. W. Recknell, wrote to a Durban newspaper in 1962, stating that he remembered, when he was at sea at the time of *Waratah*'s disappearance, hearing officers talking about their ship's dropping into an "air pocket" in the area, caused by a rift in the seabed. This was the same thing that Captain Byles called "a hole in the ocean," and Captain J. C. Brown, former master of *Pretoria Castle,* reported locating a deep fissure in the area that was clearly shown on the paper of the echo sounder.

Certainly in this area exceptionally steep seas are known. In World War II off the Pondoland coast, the cruiser H.M.S. *Birmingham* experienced a sea phenomenon similar to that which *Waratah* must have met, and the same thing was also later reported by several large British and Dutch liners.

The *Waratah* mystery was not allowed to die, and in 1972 William C. Lindemann, one of Durban's most respected engineers, considered *Waratah* was sent to the bottom as a result of an explosion in the boilers, or by a detonator left in the coal being fed unwittingly into the furnaces to set off a chain of explosions. With all this new interest, the South African papers found plenty of copy and inevitably made comparisons with *Mary Celeste*.

4. S.S. WARATAH (1909)

In March 1973 a former Royal Air Force officer, Wing Commander C. V. Beadon, who claimed to have developed a means of ascertaining *Waratah*'s position, placed her at exactly 29 degrees 46 minutes 38 seconds East, 32 degrees 36 minutes 8 seconds South, in a depth of between 21 and 24 meters. Geoffrey Jenkins, the novelist, in his book *Scend of the Sea* (1971), placed her in much the same position, and that brought another story from the 32-year-old Mayor of Grahamstown, Pat E. McGahey, who had been told years before by a senior South African Airways pilot that he had several times spotted a ship lying on her side in the position given by Beadon. McGahey went to look for himself and he claimed to have seen it on two occasions in about 30 fathoms north of the Fish River. He had also found an old laborer named Mpundulu who, remembering "a great storm" in the area years earlier, saw lightning coming from the sky while he was rounding up goats. McGahey considered the "lightning" was a distress rocket from *Waratah*.

Roos's map turned up again in 1973 in a family album, but, despite all this new activity, *Waratah* has still not been found. At the end of October and the beginning of November 1979, when the 224,000-ton Norwegian ore carrier *Berge Vanga* disappeared off Cape Town, the theories started that she, too, had dropped to the bottom of the ocean when "a sudden parting of the waves" formed a gigantic hole into which she had "slid" and then been "overwhelmed as the waves rushed back."

"This phenomenon is unique," the *Sunday Telegraph* said, "to the southern ocean, when a combination of gale, current and ocean bed contours tugs the sea apart to form an enormously deep vacuum, into which a ship can plunge and be buried by the waves."

Sailors who have survived the experience, it went on, have told of "killer waves towering high over the bridges even of super-tankers, as high as ten-storey buildings." These "holes," it stated, occurred mainly off the South African coast opposite the Transkei homeland, and it quoted the case of *Waratah*. It then went on, however, to point out that *Berge Vanga* was 600 miles due west of Cape Town—in other words, not in the same

148

area at all—but said that the "hole in the ocean phenomenon" was not being ruled out.

What was originally probably nothing more than a vivid metaphor by Captain Byles has become a theory which seems to have gained considerable ground in recent years. But a blunter note was sounded, also in 1979, in a letter to the author by George Young, shipping editor of the *Cape Times,* who had dealt with shipping in the area for 45 years. He considered that the loss of *Waratah* had been dramatized out of all recognition, and that it wasn't much of a mystery anyway. He didn't think much of "holes" or "pockets" in the ocean, but he admitted that abnormally big swells caused by a combination of current flowing south and wind blowing north sometimes caused three or four inordinately high swells to come in succession, each with a comber, or "cauliflower," on top. There had been several cases in recent years, he said, of ships' running into these abnormally high swells, and he quoted *Edinburgh Castle,* which had been making 19 knots when she hit one, and *Bencruachan,* which, while making 20 knots, struck one of these combers, the weight of water on the foredeck so pressing down the structure that the bow was in danger of breaking off and she had to return to Durban. He thought that *Waratah,* a ship without modern refinements, with wooden hatches, a speed of 13 knots and down to her marks with cargo, hit one of these seas. The load of water fell on the forward well deck and filled the holds so that she went down like a stone. The same thing had happened to other ships, he pointed out, but they had succeeded in getting a message away. Young felt there was no reason to believe the ship was topheavy or turned turtle. "She filled forward and sank," he said. As for the absence of debris, he quoted *Melanie Schulte,* a 40,000-ton German ship lost in the Atlantic which left behind her nothing but a few containers, a raft or two and a radio beacon, all of which were recovered within a few hours. He pointed out that *Waratah* had been lost some two weeks before the alarm was even raised.

He didn't even have much time for the discovery of ships lying on the ocean floor. There were about 30 of them in the area, he said, as a result of U-boat activity during World War II, and he thought that the air force pilot, Roos, had seen not *Waratah* but a Dutch passenger liner, *Colombia,* which was serving as a submarine depot ship when she was sunk in 1942 by U-156 while on a near-in course and in shallow water.

Young's theory is well supported by the events described by Recknell, by Captains Byles and Brown and Chief Officer Evans, of *Rabaul,* by the officers of *Birmingham* and other ships, even for that matter by the concern of Sawyer and Richardson with *Waratah*'s pitching. Perhaps, in fact, Sawyer was right in the end, and the disaster was not due to instability or rolling at all, but to the huge East African seas pitching onto *Waratah*'s bows so that, as Young said, she filled forward and sank at once without a chance to recover.

5. U.S.S. Cyclops (†1918)

An American tragedy

Perhaps she plies through Arctic wastes
On some dim new quest with Franklin's men
Or sees a new Pacific's blue,
As those on Darien.

This fanciful verse, from a poem by the American T. H. Ferril, which appeared in *Literary Digest* during World War I, referred surprisingly enough not to a lost exploration ship but to a vast, modern and unromantic United States fleet collier, *Cyclops,* which disappeared at the beginning of April 1918 while on passage between Barbados and the Delaware River. It was at a time when the last German push of the First World War was taking place and the Allied armies were reeling back before the pressure in the second Battle of the Somme.

Since torpedoed or mined ships were common during the First World War, it seemed likely that this was *Cyclops*'s fate. But this proved not to be the case. For Americans, the *Cyclops* mystery has exercised as great a fascination as the *Mary Celeste* mystery has for the British.

By 1918 ships normally carried radio, and it had been hoped that the days of unexplained disasters were over because a radio call should bring help long before the last survivor had drowned. But *Cyclops*'s disappearance with 304 people on board proved to be the biggest loss the United States Navy suffered in World War I. Before the storytellers and rumor mongers had finished with her, *Cyclops* had been subjected to giant squid, enemy agents, treacherous crews and even that latest bogey, Bolsheviks.

"Worse things happen on the big ships" is a sailors' saying when things go wrong, and *Cyclops* was the biggest vessel to be lost without trace and without apparent reason. She was a 10,000-tonner, one of the largest vessels afloat in those days. She was 542 feet long with a generous beam of 65 feet. She could do 14 knots and had self-trimming holds and water ballast tanks for maintaining a proper equilibrium, while 14 huge kingposts towered upward in pairs from her deck to make coaling the fleet swift and easy. Her keel was laid in 1909 at the shipyard of William Cramp and Sons, of Philadelphia, and she was launched on May 7, 1910. She was the first of three new-type ships ordered for overseas transport and the first of seven sister ships, all called after Greek mythological figures, to be built in various shipyards on the American seaboard.

The launching ceremony was performed by Mrs. Walter H. Groves, daughter-in-law of the president of the firm which built her, but *Cyclops* refused to budge until jacks were brought. For 10 minutes men struggled with her before she finally slid into the Delaware River.

Her master was a somewhat unexpected character. He was of German descent and had arrived in San Francisco from Bremen in 1878 with the name of Johann Frederik Georg Wichmann. He had gone to the States to escape his father's beatings. On arrival he was taken under the wing of a friendly retired sea captain named Worley, who took him sailing and instilled in him a love of the sea. Though Wichmann at times lapsed into his native German, he taught himself to write and speak fluent English, and his late teens and early twenties were spent on sailing ships plying between San Francisco, the Philippines and Australia. For a time in the late 1880s he ran a bar in San Francisco but then returned to sea as master of the Pacific yachts *Aloha* and *Detolina*. About 1890, when he became a naturalized American, he entered the United States Naval Overseas Transport Service, which operated colliers, transports and supply ships. These vessels were not manned by naval personnel but by any sailor picked up on the waterfront. They did not even have to be Americans. Wichmann became a warrant officer able to command tugs or other small vessels. When the Spanish-American War started in 1898, he won a lieutenant's commission, joining the transport *St. Mark* which dodged the Spanish around the Philippines until Admiral Dewey and Commodore Schley eliminated the enemy fleet.

After this he began to command colliers, his ships growing larger all the time, but always difficult, dirty ships which attracted disorderly crews. In February 1906, while captain of *Leonidas,* he married Selma Schold, of Bremerton, Washington, and though a martinet as a master, he was a good husband in the habit of writing tender letters home. Two years later the carpenter of his ship *Abarenda,* a Russian who went by the name of Dixon, murdered the third mate. Though Wichmann had no connection with the crime, the court was obviously surprised that he didn't run a more orderly ship. The incident did not affect his career, however.

In 1909, because the German Kaiser Wilhelm II was not popular with his saber-rattling and threats of war, Wichmann changed his name to George Worley. The following year he became master of the 10,000-ton *Cyclops,* which was part of the U.S. Naval Auxiliary Service. This group of ships comprised colliers, cargo, refrigerator and hospital ships, and the personnel was still civilian under the jurisdiction of the Navy Department, though shortly after the declaration of war in 1914, the ships were taken into the navy and the officers and men were controlled by the Naval Reserve.

Worley took *Cyclops* to sea for the first time in 1910 and found that, despite her technical innovations, she tended to roll and pitch badly, and, though she could carry 10,000 tons of coal, her roll was once noted at 47 degrees. But with a good crew she gave no trouble. Her chief engineer, Samuel Dowdy, was sufficiently attached to her to write an article, "Splendid Record of U.S. Naval Auxiliary Ship, *Cyclops,*" published in the magazine the *American Marine Engineer* in July 1915. She could coal a squadron of dreadnoughts and 10 destroyers for a cruise of 4000 miles, he said, and, despite her wild crews, she had already won the coaling record for the fleet—574 tons to the battleship *Arkansas* in one hour.

By this time Worley was gaining a reputation as a good master who could be tough with his crew if necessary. An Annapolis graduate doing temporary duty aboard *Cyclops* once saw him standing with a wrench in his hand forcing coal heavers, who had panicked when the ship rolled in heavy seas, to return below. But despite his strong methods he was willing to lend his men shoes, oilskins, underwear, even money, and always had a beer or a tot of whiskey ready for them from his own locker. A joke he made about keeping a lion aboard as a ship's cat was taken seriously and brought him a reprimand from the Navy Department.

Worley still thought nostalgically of Germany, even pretending that once in Kiel the Kaiser had inspected his ship. When war broke out, he made no secret of where his sympathies lay and often entertained the officers of German ships calling at ports of the still-neutral United States, particularly those from the interned raider *Prinz Eitel Friedrich*. It did not make him popular. Non-German neighbors whose sympathies were with the Allies complained of his rebuking them for defending Britain and France, and one of his yeoman, Paul Roberts, one of whose jobs was to keep Worley supplied with the latest news, noticed that he often showed his pleasure at German victories and was in the habit of drinking with the German members of his crew.

When America entered the war, however, Worley, by now a commander, appeared to become an American first and foremost, and took his ship to France with the first convoy of American troops, the entire First Division. They sailed from New York on June 14, 1917, and since the 18 transports, including *Cyclops,* were escorted by no fewer than five cruisers, 13 destroyers and one converted yacht, they were considered to be well guarded. After that, however, perhaps because of his German origins, Worley was kept on the American side of the Atlantic. There had also been complaints from his crew that he got drunk and was careless about showing lights and, according to one unlikely story, even rigged up masthead lights to signal to U-boats. It was said he had even threatened to run his ship into Rotterdam and deliver her to the Germans. The complaints came chiefly from his assistant surgeon, Dr. Burt Asper, who does not appear to have liked him, but there were plenty of others who did get on well with him.

An inquiry into Worley's behavior was held, but, though he was certainly shown to drink too much sometimes and probably showed favoritism in the matter of short leave, he did not, as was alleged, use insulting language to his crew or bully them at gunpoint. As in Britain when she first entered the war, a great deal of hostility was shown in the States to people of German ancestry. In Britain it resulted in looted shops and assaults, and it is possible that Worley was victimized for the same reason. The punishments he handed out to his crewmen were proved to be no more than those administered by other masters. Though his name was cleared, however, his ship remained in American waters, coaling naval vessels from Nova Scotia and the Carolinas, a surprising move because every ship was needed to transport men and supplies to France and *Cyclops* was one of the biggest available.

In December 1917 Worley fell, aggravating an old hernia so that he had to walk with a stick. The injury so bothered him that he told his wife he would have an operation after his next trip and hoped that it would be followed by a long leave.

Just before New Year's Day, 1918, Worley was ordered to deliver coal to Brazil and bring back for the United States Steel Products Company a cargo of manganese, used in the manufacture of arms. For this, *Cyclops* would be registered as a merchant ship under charter to the United States government and the Brazil Steamship Company. At the time Worley's daughter was sick and he was still suffering from the aggravated hernia and could have gotten out of the trip. He was in low spirits and even remarked to a friend, Samuel Harris, "This may be the last time you will see me. I may be buried at sea." Nevertheless, he made no attempt to back out and on January 8 *Cyclops* sailed from Norfolk, Virginia. She was loaded to her mark with coal, mail and miscellaneous stores for the South American Patrol Fleet, operating off the South American east coast. The day was overcast and cold, with light snow falling. For days the harbor had been completely frozen over and liberty parties from ships at anchor had even walked on the ice back to their vessels that morning.

Plowing through the ice floes in the channel, *Cyclops* narrowly averted a collision with the U.S.S. *Survey,* outward-bound for the Mediterranean for patrol and anti-submarine duties. It was to be the first of a series of untoward incidents that dogged her voyage, but by nightfall she had cleared the Virginia capes and was heading south into heavy wintry seas with a speed that was amazing for such a heavily loaded vessel. At about 8:00 P.M. Conrad A. Nervig, one of the watch officers, went forward to relieve the officer of the watch and was mystified to hear a sound rather like metal plates being rubbed together. He found the ship was "working"—twisting or bending—to such an extent that where steam or water pipes passed through web irons or were in contact with portions of the hull, the movement could be seen quite distinctly. In daylight it was even clearer. When he called the captain's attention to it later in the day, Worley shrugged it off. "Son," he said, "she'll last as long as we do."

On the fifth day out Lieutenant Forbes, the executive officer, was ordered to his room by the captain, under arrest after a disagreement about the ship's work, something, Nervig said, which was common coin on Worley's ships. The same evening Ensign Cain, one of the watch and division officers, was placed on the sick list and ordered to bed by the doctor, though apparently in good health. It was the general opinion in the wardroom that this was to save him from becoming a victim of Worley's temper. The captain made no comment.

Since it became Nervig's job to relieve Cain, he became very well acquainted with Worley. One night at midnight he was startled to see him appear dressed in long woolen underwear, a derby hat and a walking stick. He neither apologized for his attire nor even mentioned it, merely returning Nervig's salute and greeting. Nervig realized it was purely a social call.

The visit lasted about two hours and, leaning against the

bridge rail, Worley entertained Nervig with stories of his home
and the sea, most of them humorous. The nightly visits became a
habit and Nervig even began to enjoy them, but Worley's garb
never varied. Nervig considered his captain a gruff, eccentric salt
of the old school, but "a very indifferent seaman," however, and
a "poor, overly cautious navigator." Unfriendly and taciturn, he
was also generally disliked by both officers and men, though he
appeared to become quite fond of Nervig and, when Nervig received orders detaching him from *Cyclops* in Rio, even attempted
to have them revoked. Fortunately for Nervig he was unsuccessful.

As the ship came in sight of South America off Pernambuco
two weeks later, Worley changed course to take them 48 miles
past Bahia before returning to anchor off the city. "Three more
hours on that course," Nervig claimed, "would have had her
aground." The navigator had protested but had been brusquely
overruled.

There had been more troubles with the crew and Asper, the
assistant surgeon, wrote to his brother that the one thing he
wanted above all else was to leave the ship. One other strange
letter arrived home from Bahia. William Wolf, a seaman, wrote
to his mother that they were "going across from Brazil" and that
he expected to be away until the following year.

In Bahia the ship supplied coal to U.S.S. *Raleigh*, a veteran
of the Battle of Manila Bay in the Spanish-American War, but
on getting underway again twice fouled the warship, though
with only slight damage. *Cyclops* took on coal and then continued south to Rio de Janeiro, arriving on January 8. The afternoon before there had been trouble with her machinery. A
high-pressure steam cylinder on the starboard engine blew up,
and, since it could not be repaired until the ship returned to
Norfolk or Philadelphia, she would have to operate at reduced
speed on only one. An investigation was ordered by the Commander-in-Chief, Pacific Fleet, Admiral William R. Caperton.
The officer in charge, Captain George B. Bradshaw, of the
cruiser *Pittsburgh*, was startled by Worley's feeble appearance
and, believing him to be suffering from a hangover, thought him

a strange character to be a naval commander.

The inquiry did not bring up any evidence of sabotage, as was at first feared, and one of the engineer officers, Lieutenant (jg) Louis J. Fingleton, who had heard warning noises before the cylinder exploded but had taken no action, was blamed. With the rest of her cargo dumped aboard, *Cyclops* was ordered by the Navy Department's Bureau of Construction and Repair to limp home on one engine as best she could, while Bradshaw told Worley to hurry his departure.

At Rio the rest of the cargo was distributed to the various ships at anchor there, and Nervig was detached to U.S.S. *Glacier*, relieved that his tour of duty in *Cyclops* had terminated without any unpleasant incidents. For her return trip *Cyclops* took on manganese ore. There was some fear that it might capsize the ship with its weight, but *Cyclops* and her sister ships were steady vessels, it was not the hurricane season, and in her eight years of service, coal had never once shifted in a partly filled hold. Nor had her sister ship *Jason*, which had also carried manganese ore, ever reported any problems. The loading went ahead.

During this period, one of the ship's boats was tied up to the quarter boom when the engine was turned over without warning, drawing the boat into the propeller. The seaman in it was injured, knocked overboard and drowned. Nervig blamed Worley for the accident, considering that his irrational methods had thoroughly disorganized the crew.

Still concerned with his health, Worley had meanwhile met with a new problem. One of his men had killed a shipmate with a bottle in the Highlife Club. A letter to his in-laws indicated Worley's unhappiness. He was still in low spirits and clearly did not enjoy going to sea in wartime, and the letter ended with the uncharacteristic wish that the Germans would hang the Kaiser and end the fighting. His letter to his wife also complained of low spirits and of his coming operation.

At the last minute a civilian passenger, Alfred Louis Moreau Gottschalk, the 45-year-old U.S. Consul in Rio, joined the ship. He was a strange man. A New Yorker, he had been a newspaper correspondent for the *New York Herald* and the London *Tele-*

graph during the Spanish-American War, before trying sugar-planting in Santo Domingo and Haiti and eventually entering the U.S. foreign service in 1902. He was suspected of German sympathies and of exceeding his duties in cooperating with German firms in Brazil. He was also believed to have sent funds to the German Red Cross and was considered by Naval Intelligence to be an unscrupulous liar interested chiefly in furthering his own interests. He had always spent his holidays in Berlin but told his friends that when he got back to New York he intended to join the army.

Gottschalk was only part of the pro-German setup in South America. Indeed, for some time it had been part of the Pacific Fleet's task to intercept agents believed to be running between Spain and South America. As only Brazil was involved in the war on the Allied side, all the other South American ports were safe havens for German agents, wireless operators and saboteurs, two of whom had already been apprehended by an officer from *Pittsburgh*.

Cyclops sailed on Saturday, February 16, with 10,800 tons of manganese, insured for half a million dollars, in her hold. With an extra 1000 tons of coal which had been added to her bunkers, she was 15 inches lower in the water than her Plimsoll Line and drew 31 feet, as much as a battleship, but as this would counter her roll it was considered to be an advantage. In his departing cable, Worley estimated his voyage at 10 knots instead of the usual 14.

The ship arrived at Bahia on February 20, and Nervig, there aboard *Glacier,* watched her arrive—surprisingly from the north because she should have come from the south. He put it down once more to Worley's strange methods of navigation. While in the port, a friend of Nervig's from *Cyclops,* Paymaster Carrol G. Page, visited him. At the gangway as he left, he shook hands and said, "Well, good-bye, old man." It probably meant nothing, but Nervig was deeply impressed with the finality of the farewell, which proved to be prophetic.

While in Bahia *Cyclops* took on board 72 navy and Marine Corps officer passengers from U.S.S. *Raleigh,* mostly returning

to the States on leave, for discharge or for rotation of duty. There
were also several men facing disciplinary action, five of them in
the cells, three of whom had been involved in the murder at the
Highlife Club, though the murderer was not on board. Otherwise, according to one of *Raleigh*'s officers, the naval personnel
aboard *Cyclops* were sound men with good records. The ship
now had 304 people on board altogether.

She left Bahia on the evening of February 21 and *Raleigh*
signaled her estimated time of arrival at Baltimore, Maryland,
to Naval Operations as March 7. Driving through calm seas at
around 10 knots, she reached Bridgetown, Barbados, on March 3
on an unplanned and mysterious stop. An additional 600 tons of
bunker coal and 180 tons of provisions were ordered. Since
Cyclops had had 1500 tons of coal aboard when she left Brazil,
the American Consul in Barbados, Brockholst Livingston,
thought the order excessive and suggested waiting until the ship
reached Baltimore or some other east coast port, where the rates
were lower. Worley, however, seemed in a hurry and didn't even
make the customary call on the British port officer who, in a huff,
refused to go aboard *Cyclops*.

On the morning of March 4 Worley signaled his ETA and
took his ship from Bridgetown into a calm Atlantic, once again,
thanks to Worley's strange methods of navigation, steaming
south when she was supposed to be heading north. Livingston's
son noticed the phenomenon. The Lamport and Holt liner *Vestris,* on her usual run from Buenos Aires to New York, exchanged
radio messages with her 24 hours after she had left Barbados,
when *Cyclops* reported, "Weather fair" and made no mention of
any trouble or difficulty. After that there was nothing, and
neither the ship nor any of the men in her were ever seen again.

On the east coast of the United States the weather was warm and
wet and the usual spring storms, except in the north, were ab-
sent. On March 13 a young officer at naval headquarters in Norfolk, Virginia, noticed that *Cyclops,* which had been due in port,
had not turned up. He sent off a routine radio inquiry, asking her
to report her probable date of arrival. No answer was received

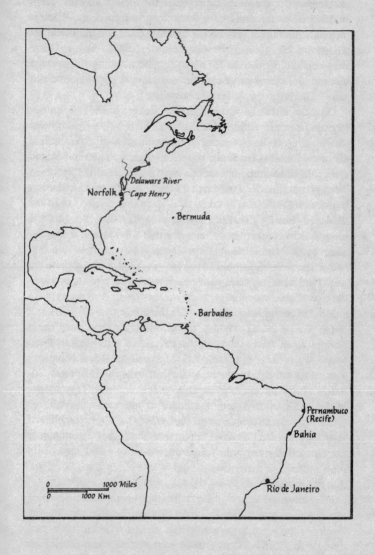

Delaware River
Norfolk • Cape Henry

• Bermuda

• Barbados

Pernambuco
(Recife)

Bahia

Rio de Janeiro

0   1000 Miles
0   1000 Km

and, curiously, the navy didn't react for almost another week. In 1945 when the cruiser *Indianapolis* was torpedoed while returning from Guam (where she had taken the atomic bomb intended for Hiroshima), it was discovered that although there were explicit instructions for naval departures and arrivals, there was no operational procedure for reporting ships which did not arrive. Doubtless the same rules existed in 1918.

On March 18, however, someone woke up and the cruiser *Pittsburgh,* which had left for the River Plate while *Cyclops* was still loading at Rio, was asked how much coal *Cyclops* had aboard when she left South American waters. The reply stated that *Raleigh* had reported that *Cyclops* was intending to proceed directly to Baltimore. On March 23 the Naval District of Charleston, South Carolina, was asked to keep calling *Cyclops* until she answered, but they replied the following day that they had not been able to contact her. They were told to continue trying because by now the alarm that *Cyclops* might be lost was growing, and if she were she would be the navy's biggest casualty of the war. That same day three patrol vessels were ordered from Guantanamo, Cuba, to search, and all stations south of Norfolk were ordered to continue trying to contact the ship.

On March 28, St. Thomas in the Virgin Islands and Consul Livingston in Barbados reported having no information. *Pittsburgh* was ordered to try to contact her and the Royal Navy was also asked to help. By the beginning of April the U.S. naval authorities were anxious and began to make inquiries. The families of the men on board were questioned about recent letters, especially from South America or Barbados. Questions were asked about Worley and German sympathizers in his crew, and whether *Cyclops* could have been seized after a mutiny and taken to a rendezvous with the enemy. Could Worley even have sailed the ship to Germany?

The routine inquiries led to the South American Shipping Company, of Manhattan, who had been responsible for shipping *Cyclops*'s cargo of manganese, and there the Intelligence men discovered a strange story that seemed to have some bearing on the case: the man who had signed the papers for the manganese

had disappeared. This man, Franz Hohenblatt, was a German
who had just taken out his first American citizenship papers.
When *Cyclops* had been dumping her coal in Rio and taking the
manganese on board, Hohenblatt in a drunken confidence had
told a fellow employee, Harry Lambert, that he hated America
because her entry into the war made it impossible for him to return to Germany. Referring to other members of the firm, he had
said, "They should be shot. They are fools for fighting for this
country." Lambert had wondered at the time whether to pass on
the confidence but, afraid that Hohenblatt might take it out on
him, had decided to do nothing. Lambert now told Intelligence
men that Hohenblatt had told him of burning 700 letters he had
received from Germany, saying, "They won't get anything on
me." What did he mean? Was he merely afraid that his American
citizenship might be held up, or had it something to do with the
missing *Cyclops?*

Then Brockholst Livingston, the American Consul in Barbados, in a long and dramatic dispatch, reported Worley's pur-
chase of 600 tons of coal when he already had enough on board
to reach Bermuda. He had asked Livingston to foot the bill on
his behalf as he did not have enough money to pay for the supplies. He had also taken on a ton of fresh meat, a ton of flour and
1000 pounds of vegetables for which $775 had been paid. Liv-
ingston also believed that more than the 600 tons asked for had
been put on board. He also referred to Worley's dislike of being
called "a damned Dutchman" by other officers and to the fact
that he had heard rumors of trouble and conspiracies on *Cyclops*
en route to Barbados, after which several men had been confined
and one even executed. Some of the men aboard, he pointed out,
were prisoners from the fleet in Brazilian waters, one of them
with a life sentence, and he had noticed that many of the crew
had German names.

The message raised a lot of questions. The epithet "Dutch"
was explained easily enough. It was a term applied generally by
English-speaking sailors to all north Europeans who went to sea
and was probably a corruption of "Deutsch," which was probably all that many could say of themselves. But why had Worley

taken on so much coal, more even than he had originally asked for, and why so many supplies? What was the trouble he had had, and what were the rumors about conspiracy and executions?

The questions troubled the navy enough for the Chief of Naval Operations, Admiral William S. Benson, to suggest in a message to Admiral William H. Sims, Senior American Naval Officer in Europe, that perhaps what had happened to *Cyclops* was not a sinking. But he admitted that this view was probably based on the instinctive dislike Livingston had clearly felt for Worley. Naval Intelligence believed that the disappearance was probably due to a magazine explosion, and it was felt she was not sunk by enemy action, as the Germans usually announced the destruction of large ships and no such announcement had been made for *Cyclops*. Still no serious search had been made by April 15, by which time *Cyclops* was a month overdue.

On that day alongside the news of the Second Battle of the Somme, the *New York Herald* carried the headline "Big War Supply Ship Vanishes Without Trace." The *Virginian-Pilot and Norfolk Landmark* announced only, "American Naval Collier Probably Lost: Fate May Be Another Mystery of the Sea." Devoting a column of space to the disappearance and the names of Norfolk men who were aboard, it announced that *Cyclops* had been overdue since March 13 and that there was extreme anxiety about her safety. Despite this, the navy, though admitting she was missing, let it be known that there was no good reason for her being overdue. The statement said,

> No well-founded reason can be given to explain *Cyclops'* being overdue, as no radio communication with or trace of her has been had since leaving the West Indies port. The weather in the area in which the vessel must have passed has not been bad and could hardly have given the *Cyclops* trouble. While a raider or a submarine could be responsible for her loss, there have been no reports that would indicate the presence of either in the locality. . . .

It went on to say that although one of *Cyclops*'s two engines was damaged and she was proceeding at a reduced speed, even with both engines disabled she could still use her radio. The navy was giving nothing away. The reporter, Harry P. Moore, who had first discovered the story, found that Washington officials not only belittled his story but even threatened proceedings if he continued to insist that the ship was missing.

The day after the navy's statement appeared, the *Virginian-Pilot* announced that officials in Washington refused to believe that the great collier and the 304 people on board could have been wiped out without leaving a trace. It was stated that orders had gone out for searching vessels to quarter every inch of the route covered by *Cyclops* and to visit "every one of the scores of islands that dotted that portion of the sea." Some officials were admitting now that not one of the theories advanced so far to explain the disappearance seemed plausible in the face of the known facts. Though an internal explosion might have destroyed the radio and engines at one go, surface wreckage would surely have remained to mark the ship's grave. No one seemed to have heard of *Waratah*. The statement went on to say that "a sudden hurricane, not infrequent in those waters," might have disabled and engulfed the collier, but that in this case some evidence of the disaster must surely have been left.

The search was continued vainly until May. The navy calculated the ship's stability and decided that she would have had an uncomfortable, quick but not excessive roll. It also admitted that it was possible for the heavy ore to shift enough to cause a list and submerge the edge of the deck and that such a shift would have been dangerous under certain conditions of sea and weather, but—once more—that no unusual weather had been encountered by other vessels along the same route.

Naval Intelligence drew up a list of theories to explain the loss of the ship: mutiny in the crew, who had then left the usual trade route; Gottschalk might have arranged to hand over the ship to the Germans; the ship might have been torpedoed; the cargo of manganese dioxide, highly incendiary under certain conditions, might have exploded; the ship might have turned turtle or broken up because of excessive rolling; Worley might have surrendered the ship to the Germans or connived at her destruction by submarines. Nevertheless, the navy said coldly, there was no evidence to support any of these theories and the Navy Board in Rio refused to contemplate sabotage aboard—which, however, would not have been all that difficult.

The rumors continued, the most prominent of which were that Worley had scuttled his ship or sailed her for Germany. This story seemed ridiculous because it would have meant that Worley —alone or with a few supporters—had managed to overcome the biggest part of the 304 men on board—most of them honestly engaged in the war on America's side against Germany. Yet a few days after the signing of the Armistice, Admiral Benson signaled Admiral Sims: "Make every effort to obtain information from German sources regarding disappearance of U.S.S. *Cyclops*."

The story upset Worley's wife, who insisted her husband was a good American. She added that the usual unidentified informant had told her *Cyclops* was safe. But where? Unable to face neighbors and investigators, she left Norfolk on May 7 for Port Orchard. On June 1 *Cyclops* was declared lost and all aboard her legally dead. By the end of the year Admiral Sims had discovered that neither German U-boats nor German mines could have been responsible for the disaster.

Though *Cyclops* was now officially regarded as sunk, the rumors continued: She had been torpedoed by a Brazilian destroyer. She had been seen at Kiel, Antwerp, in the Adriatic, off Colombia, in Antarctica. Russian Bolsheviks, who had become the bogeys of the Western world since the revolution, were suggested. Other stories suggested that *Cyclops* had gone aground or been wrecked on some deserted, uncharted island in the Caribbean and the 304 men were still waiting to be rescued.

Many supposed clues appeared, but none was authentic. In 1919 the mother of one of the crew members received a telegram from New York saying her son was safe and that *Cyclops* was in a German port. It was found to be a hoax. The usual bottles were washed ashore along the eastern coast of the United States. One found near Valesco, Texas, said *Cyclops* had been torpedoed on April 7, 1000 miles off the coast of Newfoundland. This was also found to be a hoax like all the others, in which the names represented no one on board *Cyclops*. Planking, life belts, an entire barnacled wreck in the Bahamas were investigated, and in 1920 a destroyer was sent to the Bahamas to search for what was alleged to be the hull of *Cyclops*. It was even suggested that a disappearing "submarine fortress" had blown *Cyclops* out of the water with 15-inch shells, and a reputable magazine seized on the story and improved on it. The *National Marine* published a story that she had been crushed to matchwood and dragged under by a gargantuan squid. Gargantuan it must have been to have taken down the 10,000-ton *Cyclops*.

An astrologer offered his services and T. H. Ferril produced the lengthy poem in which he suggested *Cyclops* had joined Franklin's ships in the Arctic.

There was only one clear fact: after her departure from Barbados, not a single ship had sighted *Cyclops* on her way to Baltimore. The suggestions that Worley had taken the big collier to Germany were silly, because, above all else, he could never have dodged the British and American blockading patrols. He was clearly a sick man and, judging by his last letters, probably troubled, too, but this surely could not have caused the disappearance of his ship.

One of the most popular speculations was that of a sudden shift of the heavy cargo which had caused the ship to turn turtle. Commander Mahlon S. Tisdale, who had once served in *Cyclops*, suggested this in a paper to the U.S. Naval Institute in 1920. He had noticed that her sister ship, *Neptune*, in which he had also served, had a tendency to develop a sudden list, something he attributed to the design of her water tanks, which were rigged in such a way that hundreds of tons of water could slosh from one side of the ship to the other.

He described his experiences in *Cyclops* in rough weather during an exercise off the New England coast when the ship had been light. He had been thrown by a lurch of the ship from a runway, which covered the steam pipes for the winches and was used as a flying bridge, onto the deck and, landing near a topside manhole tank, he had noticed that the plate was not secured.

Alarmed, he informed the captain, but Worley had laughed at his earnestness and said they were always left off in accordance with navy yard instructions. Tisdale now suggested that the cargo might have shifted and allowed the sea to rush into the open tanks, causing the ship to capsize, something, he felt, that could occur in seconds and before she could be abandoned. With everything secured for sea there would be little wreckage, no debris and no time for an SOS, and any small gear that floated off would be dispersed long before a search started.

This was a universally accepted solution for a long time, but it assumed heavy seas, whereas *Cyclops*'s radio message to *Vestris* had stated that the weather was fair. And Commander I. I. Yates, of the Norfolk Navy Yard, pointed out in his reply that when Tisdale had discovered the open manholes *Cyclops* had been in a light condition, so that it would be immaterial whether they were on or off and that in this condition the topside tanks which the manholes served were generally kept full of water anyway. There were no orders from the navy yard about leaving them off, he went on, and he suspected that Worley, who was known as a joker, was pulling the young Tisdale's leg and would never have sailed without them in place when the ship was loaded. Certainly, in his report, Tisdale had admitted that though the ship was "cavorting round the ocean like a frisky colt," she was noticeably taking no seas over the main deck. He had noticed, however, that the clinometer had registered 48 degrees to port and 56 to starboard, and he wondered if perhaps the manganese had shifted and contributed to the list. Perhaps, he thought, even the one sound engine had failed or the steering mechanism had jammed and, without the starboard propeller to compensate, *Cyclops* had heeled over, causing the manganese to fall to one side to give the ship a list from which she could not recover. The navy's Bureau of Construction and Repair could not agree. Deep-loaded as she was, *Cyclops* should have been stable and did not take in water in rough seas.

Then a Captain Zearfoss joined the argument, deciding that *Cyclops* had been sunk by her own cargo. He explained how manganese had a tendency to settle down, grinding away whatever was below it, and thought the ship's bottom had simply dropped out.

Many naval men thought *Cyclops*'s topheavy superstructure was to blame. Her massive steel sampson posts and derricks, designed for the rapid loading and discharging of coal, towered high above the deck, and they considered that if she listed too severely the topheavy equipment would have slowed recovery, causing the vessel to capsize. Partially filled holds, such as *Cyclops* had, would be more likely to cause the cargo to shift than full holds. The flaw in this theory seemed to be that it would work only in bad weather and the facts seemed to show there had been none.

The one thing that seemed quite clear was that *Cyclops,* whatever had happened, had been overwhelmed at incredible speed, even in seconds. No SOS was received and not even the smallest fragment of wreckage was identified as coming from her. Francis S. Gibson, a lieutenant in *Raleigh,* felt that she must have "hit something at the wrong angle" and gone down within 10 seconds.

For a time, significance was attached to a requiem notice in the Rio newspapers for Consul Gottschalk, which was supposed to have been published before the news of the ship's loss had been released, but in fact it turned out to have been published several days afterwards. Curiously, the atmosphere of espionage and sabotage that existed in Rio when the ship had been there was never mentioned, and Naval Intelligence did not seem to recall the fact that the Pacific Fleet, too small to be a true fighting force, had been designed to protect United States ships and shipping lanes and those of Brazil from enemy intrigue. The two agents intercepted by *Pittsburgh* at this time also were overlooked.

Yet, as was pointed out, the Germans were experts at sabotage. Their agents had destroyed Black Tom Pier in New York in 1916 and millions of dollars' worth of ammunition, the dock, warehouses, barges, ships and trains had been blown sky-high. Plotters led by Franz von Papen, later Nazi Minister to Turkey and one of the men arraigned at Nuremberg, who was then military attaché at the German Embassy, had manufactured incendiary bombs aboard the interned *Friedrich der Grosse* at Hoboken. Among the group was also the famous Franz von Rintelen, and their devices had caused fires on 40 outwardbound ships. And what of Franz Hohenblatt, who had arranged the cargo of manganese and later disappeared after destroying his letters from Germany? Could he have been connected with the group aboard *Friedrich der Grosse* or connected with German agents in Rio?

According to the official findings (though never the view of the writer K. C. Barnaby) the Italian navy had lost the warships *Leonardo da Vinci* and *Benedetto Brin to* Austrian saboteurs, and the British had lost *Vanguard* and *Natal,* both of which cases were attributed at the time to enemy agents. Even a plot to dynamite the ways at a great emergency shipbuilding yard in the Delaware River had been discovered shortly before *Cyclops* vanished. But still enemy sabotage was not considered.

Cyclops was the first steam-propelled U.S. Navy vessel to disappear with no clue to her fate. It was felt she couldn't possibly have broken up in a storm because, despite her tenderness, she had proved herself a rugged ship over eight years of battling with Atlantic gales, so that there seemed no reason why she should turn turtle. Not only was she deep enough to be extra stable, but she also had stabilizing tanks built into her. Besides, as was pointed out again and again, there had been no storm.

Dr. H. D. Stailey, of California, who had been a lieutenant in *Raleigh* in Bahia, felt that an internal explosion from a bomb must have been the cause because of the absence of radio messages. It had to be something sudden and violent, he felt, because she could hardly have sunk without being able to send an SOS. Josephus Daniels, Secretary of the Navy, wrote *Cyclops*'s epitaph. "There is no more baffling mystery in the annals of the Navy," he said, "than the disappearance of the U.S.S. *Cyclops.*" Other navy officials echoed him with the view that . . . "the disappearance of *Cyclops* is as baffling a mystery today as it was in March, 1918," and the case was summed up by President Woodrow Wilson's pontifical "Only God and the sea know what happened to the great ship."

Every March, the anniversary of *Cyclops*'s disappearance resurrected the story again, and in 1930 she came into the news when parts of a diary sent to the navy "revealed" that four men in enemy pay had placed dynamite around the engine and sunk the ship. At first it was felt the diary might be genuine, but the belief soon disappeared when the diary went on to describe how an enemy ship with a crew of 700 was standing by to clear the sea of debris.

As late as 1956 it was reported that *Cyclops* was seen to blow up in the Straits of Florida "just before Easter," 1918, but there was no good explanation of why this report had been held over for almost 40 years or why the ship was so far off course several weeks after she was overdue.

In 1969 Nervig, the watch officer who had been the recipient of Worley's nighttime visits in his underwear, wrote an account for the United States Naval Institute *Proceedings* of what happened while he was on board from a diary he had kept, and suggested that the ship had broken in half. He mentioned the fact that he had discovered that the ship was "working" to the extent that where pipes were in contact with portions of the hull the movement could be distinctly seen. The deck amidships, he continued, was rising and falling as if the ship "were conforming to the contour of the sea." He didn't believe *Cyclops*'s cargo had shifted, causing her to roll over, because manganese ore would shift very little if at all. Though he had no idea how the ore had been loaded, he assumed that with the executive officer confined to his cabin under arrest it would have been supervised by a young, inexperienced officer. He, Nervig thought, would have distributed it in the midships holds instead of along the length of the ship, so that the ship broke up under the stress, the two sections standing on end before they sank. Because of this no lifeboats were launched and no SOS was sent.

This kind of faulty loading of a heavy cargo was suggested by the German court of inquiry as the reason for the disappearance—almost without trace—of the motor ship *Melanie Schulte* in 1952. But nothing could be proved, and in *Cyclops*'s case the ship had been loaded under the supervision of Captain Worley—who would surely have known his own ship after eight years—and of Manuel Pereira, the foreman of the Brazilian Coaling Company. Pereira, who had been in charge of loading ships for many years, stated flatly that the ship could have carried 2000 tons more ore without being endangered. The cargo, he said, was well trimmed throughout the ship.

The route *Cyclops* should have taken after leaving Barbados was one of the ocean lanes where steamships were frequent, and it seems improbable that if she sent out a radio message it would not have been picked up. Though she had a small magazine forward and another aft, there was little ammunition in either, and even if the magazines had exploded, the ship ought to have remained afloat long enough to send distress signals. And though manganese mixed with coal dust becomes inflammable under certain conditions, it was always doubted that a quantity of explosive large enough to sink the vessel within seconds could have been introduced into the cargo by agents.

None of the more normal theories was totally implausible, however, because cargoes do shift and radios do break down, and Worley had been seen to threaten with a wrench men who had been frightened by the ship's rolling. But what was the reason for Worley's remark to his friend, Samuel Harris, that it would be the last time he would see him? What was the meaning of William Wolf's letter to his mother stating they were "going across from Brazil" and that he expected to be away until the following year? With a name like Wolf, he could have been of German descent like Worley, so could there possibly have been something in the theory that Worley had kidnapped his ship? Missing ships did later turn up in Russian ports as, in more recent years, ships missing in Asian waters have turned up in ports of North Vietnam.

For years the speculation remained subdued, then in 1972 a new theory emerged, strangely enough from a man who was not even particularly looking into the *Cyclops* mystery, but into another one, the Bermuda Triangle mystery.

This area of the sea, bounded by Florida, Puerto Rico and Bermuda, became news when journalists noticed that it appeared to have seen the last of not only a number of airplanes but also a number of ships. The mystery seems to have started on the afternoon of December 5, 1945, in a strange aviation drama involving five navy Grumman Avenger torpedo bombers carrying 27 men. They vanished in clear weather, after radioing that they did not know in which direction they were flying. A Martin Mariner flying boat with a crew of 13 sent to look for them also vanished, and the naval authorities admitted after a lengthy investigation that they were more confused than before the inquiries began, and concluded that they "were not able to even venture a good guess as to what had happened."

In 1947 an American superfortress vanished in the same area, then between January 1948 and June 1950 two British airliners, *Star Tiger* and *Star Ariel,* a DC-3 and a Globemaster.

As time went on there were more disappearances—airplanes of all shapes, sizes and styles, and vessels from small yachts through large merchantmen to nuclear submarines, of which two were lost in the area. The strange part of the story was that neither wreckage nor bodies were ever found, and according to the stories there were some strange phenomena involved such as compasses spinning wildly, crooked radio beams, or violent disturbances during calm weather.

Enterprising journalists recalled an off-the-cuff comment by a naval officer involved in the inquiry into the loss of the Avengers —"They vanished as completely as if they'd flown to Mars"—and the legend was born. The incidents were still occurring and, digging into the records, they found other incidents involving airplanes and seagoing vessels. Even Joshua Slocum, the world's first solo circumnavigator, had vanished with his sloop *Spray* while crossing the Triangle. Digging further, the journalists noted that even Columbus's men had been worried by the area, and the last century had provided a whole list of vessels which had vanished or been found abandoned there.

The area received much attention, and was the subject of books, magazine articles and radio and television shows. There were other equally dangerous areas of the ocean—the North Sea, the Bay of Biscay, the area north of Iceland—but they were ignored. Television specials were devoted to the Bermuda Triangle, it figured in unidentified flying object and ancient astronautical mysteries, and according to all accounts something very odd was going on there.

Then in 1972 a levelheaded reference librarian at Arizona State University, Lawrence Kusche, became involved. He was frequently asked to help in finding information on the subject, and, while compiling a bibliography for anyone who might be interested, he decided to investigate the mystery himself and try to find rational explanations for the disappearances. Like the others he examined the records as far back as Columbus and in greater detail from 1840, and found that of the ships supposed to have been lost or abandoned in the Bermuda Triangle or just after leaving it—*Mary Celeste* was one—in most cases there was a perfectly sound reason for their disappearance, even that some ships supposed to have been in the area were nowhere near it. Among the ships he noticed that were supposed to have been lost in the Triangle was the collier *Cyclops.* Investigating a little more closely, he discovered that in the past few years there had been an important new development in the mystery.

In June 1968 the U.S.S. *Pargo,* while searching for the lost nuclear submarine *Scorpion,* located a wreck 70 miles off Cape Henry, Virginia, in about 180 feet of water. *Pargo* lowered two two-man diving teams who found the wreck lying at an angle of about 40 degrees, but, because they had landed on the extreme tip of the bow, none of them was able to give a positive identification. U.S.S. *Kittiwake* was called in the following year to investigate, and Senior Damage Controlman Dean D. Hawes, a master diver, went down to check.

The ship, clearly an old cargo vessel, startled him by its strange design. Its bridge was situated high above the deck, supported on steel stilts, and she had two cradled booms forward and what appeared to be two kingposts, with no rigging of any kind on them, and no catwalk between them. He made three dives and realized that a line of kingposts, upright beams, ran almost the entire length of the ship "like the skeleton of a skyscraper." The ship dated, he felt, from the early 1900s and gave the appearance of having broken in two somewhere aft of the first set of uprights and the bridge. *Pargo* had found another piece of wreckage nearby which could have been the stern section lying on its side. After his third dive, Hawes was forced to surface by bad weather and the ship had to leave the area.

The divers gave their impressions to a submarine force artist who drew a composite picture of the ship, and later, when Hawes, quite by accident, saw a picture of *Cyclops* for the first time, he was convinced it was the ship on which he had stood. The position had been pinpointed by some of the most sophisticated navigation gear then available at exactly latitude 37 degrees 26 minutes N, longitude 74 degrees 41 minutes W, bang on the course *Cyclops* would have followed on its journey to Chesapeake Bay. With the assistance of Representative G. William Whitehurst, of Virginia, Hawes managed to persuade navy officials to investigate and divers were sent to attempt to find the wreck as part of a training exercise.

At this time, the librarian Kusche was involved in his own research and the announcement of the discovery of the wreck caused him to wonder. If the wreck was *Cyclops*, how did it come to sink so close to its destination and why had it not sunk earlier on the voyage?

He then worked out that, with its disabled engine, its restricted speed would be only 10 knots or about 240 nautical miles a day. The ship would have followed the 15-miles-a-day North Equatorial Current for about 1300 miles until it met the Gulf Stream, which would then carry it as much as 120 miles a day for the remaining 500 miles. Based on these estimates, he decided that after a little more than six days at sea, *Cyclops* could have been where the sunken ship examined by Hawes was discovered. Since she sailed from Barbados in the early evening of March 4 she should have been approaching Norfolk on the night of March 10. Like Sir William Crocker and Dr. Cobb with *Mary Celeste,* and Admiral Rickover with *Maine,* Kusche ignored the legends and examined the details of the available facts. Convinced that only a storm could have caused *Cyclops* to disappear, he set out to find one, despite the fact that *Cyclops* was said to have had fair weather.

Since a north wind has a reputation for raising havoc with the Gulf Stream, he felt the storm he was seeking must have come from that direction. A north wind strikes violently, he found, against the opposing flow of the current, turning it into a wild torrent that has overwhelmed more than one ship. If he could find such a storm the wreck might well prove to be that of *Cyclops*.

Kusche obtained east coast weather reports for the period. He soon found that the navy's statement after *Cyclops* had been found to be overdue—that there had been no storms—was not quite correct. True enough, there had been nothing to trouble her near the West Indies where they had been searching, but by that time, Kusche decided, she was no longer there but off the coast of Virginia where the Navy Weather Bureau had found a front lying right across the position Kusche estimated for her. The records showed that early in March 1918 the wind blew hard in Norfolk, reaching a top speed of 30 to 40 knots almost every day. On the 8th it died almost completely, but began to build up again on the morning of the 9th, when gale warnings were issued from Maine to North Carolina. Shifting steadily from the southwest, the gale increased in strength all the time until by 10:00 A.M. on the 10th—just about the time Kusche estimated *Cyclops* would have been off Cape Henry near the wreck Hawes had seen—it had reached 58 miles an hour. After noon the wind had shifted again, blowing from the northwest at 60 miles an hour. The speed had varied between 40 and 60 until 5:00 P.M., then had remained near 40 for the rest of the evening, finally tailing off about midnight.

The storm had been widespread. Peak winds of 84 miles an hour struck New York City, where they caused one death, and gale warnings had been extended as far south as Florida. The steamer *Amolco,* 375 miles northeast of Norfolk, was caught in the storm from noon on the 9th until the afternoon of the 11th, and had to heave to for the entire two days. Her commanding officer considered it the worst storm he had ever encountered. The seas had strained the hull and machinery, stove in the lifeboats and wrecked the bridge, causing $150,000 in damage. Mate W. J. Riley later told the *Virginian-Pilot* that *Cyclops* was probably caught in the teeth of the same gale, and he was positive that she sank "during the raging of the high winds." Several seamen, aware of *Cyclops*'s tendency to roll, agreed and added that she must have sunk before lifeboats could be lowered.

The navy's statements after *Cyclops* disappeared all suggested that she had been lost near the West Indies, and the order to ships to visit every one of the islands that dot that part of the sea and the suggestion that a hurricane, "not infrequent in those waters," might have been the cause, indicated that they were looking so far away they had not realized the disaster might have happened under their very noses. The absence of any radio messages after the one to *Vestris* strengthened the idea.

In Norfolk strong winds are not uncommon and the storm from the 9th to the 11th did not receive much publicity. The *Virginian-Pilot* mentioned it only in a half-inch paragraph on March 10 at the bottom of page two, saying only that gale warnings had been issued at five the previous afternoon from Maine to North Carolina. There were no references to the storm in following editions.

By the time *Cyclops*'s disappearance was reported, Kusche decided, the storm had been tucked away in the weather bureau's statistics sheets and forgotten until he found it 56 years later. At the time neither the weather bureau nor the newspapers had drawn the navy's attention to it, and *Amolco* never appeared again in the press after the one small article giving Mate Riley's views on *Cyclops*. Because of the war, there was never an official inquiry. Had there been one, the weather would surely have been noticed.

Up to 1974, when Kusche's book went to print, it seemed that perhaps he might have hit on the answer to the mystery, especially when, in July that year, the minesweeper *Exploit* took up the search, armed with sophisticated navigation gear and ultramodern sonar equipment, and quickly located the wreck. The

spot was marked with a buoy, and Lieutenant Douglas Armstrong, the captain, announced to the *Virginian-Pilot*, the paper
which had first reported the loss of *Cyclops*, that the long-lost
ship had finally been found. There was considerable activity in
Norfolk and Hawes watched as underwater television cameras
examined the ship. He was startled because it was not the ship he
had found. Though the ship was not identified, it was found to be
filled with scrap iron and was equipped with radar, which was not
even thought of when *Cyclops* was lost. Armstrong seemed to be
wrong and Kusche's theory seemed to have been exploded, and
in August 1974 the submarine rescue ship *Opportune* finally determined that the sunken wreck was not *Cyclops*. The best guess
was that it was *Ethel C.*, a 329-foot Greek freighter that sank on
April 15, 1960, with no loss of life.

It was a bewildering turn of events. It had seemed quite certain that *Cyclops* had been found at last, but now she seemed to
have disappeared again. The possibility of the wreck's shifting
position was not even considered and for three days divers
combed the area where the wreck had been pinpointed two years
before, but failed to find any sign of it. In the navy's view, the
disappearance of *Cyclops* remained just as much a puzzle as ever.

But there was another amateur detective in the field, and perhaps the soundest theory of all was that put forward by Rear
Admiral George van Deurs, USN (ret). His article on *Cyclops*
in the Naval Institute *Proceedings* in January 1970 had gone unnoticed by the newspapers and by Kusche, and in October 1974
he expanded his views in the *Naval Engineers' Journal*. Before
the cold technicalities of the expert, all the romantic theories disappeared once more.

Like Kusche, Admiral van Deurs had originally really been interested in something else entirely—what had happened to *Cyclops*'s sister ships, *Nereus, Proteus, Jason, Orion, Jupiter* and
Neptune. *Jupiter*, he found, was converted into the United States'
first aircraft carrier, *Langley*. *Neptune* was laid up in 1922 when
warships converted to oil and was sold for scrap in 1939. On
December 2, 1925, *Orion*, loaded with coal, left Newport News,
Virginia, for Newport, Rhode Island, and was caught in an At-

lantic storm. An extra big roller dumped her on the Chesapeake Bar, buckling her amidships. She managed to limp to Norfolk Navy Yard, where the report on her twisted hull revived interest in the *Cyclops* mystery. Secretary of the Navy Curtis Wilbur thought her damage could explain the loss of *Cyclops,* but the Board of Inspection and Survey reported that her broken back was entirely the result of hitting the bar. She was finally broken up in 1933.

Admiral van Deurs had himself served in another of the sister ships, *Jason,* which also ended in a scrapyard. *Nereus* and *Proteus* rusted away in the Norfolk Navy Yard until 1941, when the wartime need for shipping caused them to be sold to a Canadian firm. They were refitted, manned by small crews and employed carrying bauxite from the Virgin Islands to Canada. Like *Cyclops,* they both vanished without trace within a few weeks of each other following much the same route, *Proteus* in November 1941, *Nereus* in December. The mysterious disappearance of these ships caused little stir in the newspaper offices because America was more concerned at the time with the Japanese attack on Pearl Harbor. Both were assumed to have been torpedoed because German submarines were active off the American coast.

In 1970 Admiral van Deurs took a different view, because he had noticed something which was common to the three ships and possibly to a fourth. As in the case of *Cyclops,* no wreckage was ever found, and research showed that at the time of their disappearance none of the three was involved in violent storms, though they were all heading into areas where winds of 30 to 40 knots were expected, with 16- to 18-foot head seas. Such waves, lower than the ships' sterns, could hardly have capsized the ships so quickly as to leave no wreckage, but van Deurs had noticed while serving in *Jason* from 1929 to 1932 that on the well deck or the bridge the ship felt like any other ship, but aft, even while moving through a glassy sea, expansion joints between her towers clattered and her stern, jumping a foot or more, kept time to the engines. Walking was like balancing on a springboard and being joggled at the same time by someone unseen.

As with most merchantmen of the time, builders constructed

the ships of metal plating heavy enough to rust for years without becoming dangerously thin. Unlike most ships, no fore and aft bulkheads stiffened their cargo spaces, but instead several heavy I-beams with three-foot webs and 10-inch flanges ran the length of the ship just inside the skin. For some 50 feet, between the engine room and Number 10 hold, the girders were normally buried in the coal in the bunkers surrounding the fire room.

In 1932, as she was being prepared for the voyage home, a seaman tapping rust from *Jason's* waterline suddenly lost his hammer through a hole he had knocked in her side—something that happened to the author on another ship in 1939. At about the same time, the first lieutenant, Charlie Allen, went into her almost empty fire room bunker at a time when the sun was flooding in. The inside of this hold, he saw, looked different from the other cargo holds and he suddenly realized there were none of the I-beams he had expected to see, only edges of rust to show where they belonged. At last he understood *Jason's* bouncing. Sulfurous coal had corroded away completely the stiffening girders in the bunker. For some 50 feet of her length, only rust-thinned skin plates attached *Jason's* engines to the rest of the ship.

When the report of the ship's condition reached the Commander-in-Chief, Asiatic Fleet, he canceled *Jason's* projected voyage back to the States and directed her via a longer, smoother route, and, lightly laden, she took two months to squirm home across the Pacific to end in a scrapyard, surviving the voyage, Admiral van Deurs insisted, simply because she had never had to buck head seas with a full load of ore.

He pointed out that a lot was learned about ocean waves after 1941, and, though *Cyclops, Nereus* and *Proteus* had not encountered any severe storms, they had each passed through a moderate cold front with winds of 30 to 40 knots, though these were not strong enough to bother 20,000-ton vessels. However, the U.S. Navy Oceanographic Office pointed out that these cold fronts were moving fast enough to outrun the waves the winds kicked up so that without warning the ships met head winds followed by high seas. In each case the waves would be roughly half a ship's length apart so that, with the load concentrated amidships, the ships were being supported by waves at each end but unsupported and sagging in the trough in the middle. Did they, like *Jason,* have a weak point that might have failed under such strains? Admiral van Deurs thought they did.

Nereus and *Proteus* had rusted even longer than *Jason,* and *Cyclops*'s unbalanced single engine had been kicking her stern up and down for hundreds of miles. All three heavily laden ships, he felt, probably snapped at their fire rooms as they bucked hard seas. If this happened, he said, their end could have been as sudden as that of *Chuky,* a 7000-ton, five-year-old Glasgow-built freighter similar in shape but 100 feet shorter than *Cyclops.* In 1926 her sister ship *Toco,* bound for Japan with copper wire, had disappeared without trace soon after reporting good weather. Some months later, also in good weather with a cargo of copper concentrate, *Chuky* reached the same area on a similar voyage. After a morning chat with her captain on the bridge the chief engineer walked aft on the catwalk. As he stepped onto the poop, a big wave rolling sternward snapped the ship in two 100 feet behind his heels. As he spun around, he saw the bow section flip over and sink so quickly that even the men on the open bridge were carried down with it. The rear section remained afloat only long enough for the chief and his men to lower a boat and get clear with the ship's dog.

It was van Deurs's opinion that *Nereus, Proteus* and *Cyclops,* their I-strengtheners corroded to nothing, all snapped in two at the fire room and sank as suddenly as *Chuky.* He pointed out that as *Cyclops*'s coal barge came alongside at Bridgetown, Barbados, with the disputed coal, her well-deck crew lowered two of the long booms and took the wooden cover from the bunker hatch between the galley door and the break of the poop. Two men filled the bunker from the coal barge alongside, one standing at the rail to signal and the other handling the steam winch. But no one went into the bunker, whose A-shaped bottom straddled the boilers so that when the bucket dropped its one-ton bite through the hatch the coal slid down to fill the space around the fire room and nothing amiss was seen. Then, for two uneventful weeks, the ship had bounced through gentle tropic seas, with one of its three cylinders disconnected, a standard engineering expedient to bring a limping steamer home. Because of this, no one worried about the engine's unbalanced thump, which bounced the after third of the ship so much that the men walked as if on a jiggled springboard, though it was almost imperceptible on the bridge or well deck. As Nervig had noticed, the ship had always "worked," just as *Jason* "worked," and, with *Orion*'s broken back probably the result of more than being dumped on the Chesapeake Bar, Admiral van Deurs's story confirmed what Nervig had seen and what Kusche had worked out. Wherever *Cyclops* lies, in the Caribbean or off Norfolk, her end probably came as suddenly as *Chuky*'s, and she broke in two during a storm. It seems more than likely.

6. M.V. Joyita (†1955)

Mystery in the Pacific

Ships continued to go missing after *Cyclops*. Oddly enough one of them was *Vestris* in 1928, the Lamport and Holt liner which had been the last to contact *Cyclops* by radio. She was another tender ship and when her list grew worse in a storm, her master, Captain W. J. Carey, who did not survive, seems to have panicked enough to make several bad mistakes, and fewer than half of her passengers were saved.

More ships followed. The list was a long one and there were the usual number which remained unexplained: the steamer *Asiatic Prince* and the five-masted bark *København* (1928); the tanker *La Crescenta* (1934); the steamers *Vardulia* (1935) and *Haida* (1937); the tramp *Anglo-Australian* and the four-masted training ship *Admiral Karpfanger* (1938); the steamer *Novadoc* (1947); the Liberty ship *Samkey* and the tramp *Hopestar* (1948); the submarine *Affray* and the 20,000-ton Brazilian battleship *Sao Paolo*, while under tow to a breaker's yard (1951); the tramp *Pennsylvania*, the steamer *Edna*, the brand-new *Melanie Schulte* and five Norwegian sealers (1952); the *Yewvalley* and eight assorted small ships (1953); the huge Argentine steamer *General San Martin*, the converted LST *Southern Districts* and six French fishing vessels (1954). So the list goes on and, according to Lloyds' Register, 1978 was the worst year ever, with a total number of 473 ships lost and a total tonnage of nearly two million. Yet none of these ships was lost by enemy action during a war. In some cases there were suspicions of age, overloading, tenderness or bad weather, but in one or two the ships were new and bad weather was not reported. Waterspouts, meteorites and pirates put in their usual appearances, but in every case there was no real clue, no wreckage and no survivors.

But for sheer mystery perhaps the most perplexing case was that of the twin-screw 70-ton motor vessel *Joyita,* in the Pacific, which bade fair to become a modern-day *Mary Celeste.* Like *Mary Celeste,* she was small and unimportant, her crew and passengers were few in number and she was found just about a month after she sailed, derelict, with no sign of her people and, stranger still, her cargo gone, too. As with *Mary Celeste,* there have been many explanations, some of them much the same as were offered for *Mary Celeste,* but, like *Mary Celeste* again, not one of those on board ever reappeared and the mystery has never been solved.

Joyita left Apia, Samoa, on October 3, 1955, for the Tokelau Islands with 25 people aboard. Despite an intensive air and sea search, she vanished from human ken for 36 days until she was spotted, waterlogged, deserted and drifting, 167 miles from the island of Vanua Levu in the Fiji group, hundreds of miles from where she should have been.

The passengers had included three New Zealand government officials, two European employees of a well-known copra trading company, Gilbert and Tokelau islanders, and her master, Captain Thomas Henry Miller, a lieutenant-commander in the Royal Naval Volunteer Reserve and a highly experienced seaman. His crew included a bosun and engineer who had sailed with him before.

The mystery was even more puzzling because there was no reason for anyone to leave the ship since, as she had been made into a refrigerator vessel for carrying frozen food cargoes, there were 640 cubic feet of cork in her hold, making her virtually unsinkable. Since she carried no lifeboats, why had 25 people taken a chance in shark-infested seas in carley rafts, when it was clearly safer to remain aboard her?

Joyita was constructed in 1931 by the Wilmington Boat Works in Los Angeles as an oceangoing yacht. Her sturdy hull was of heavy two-inch cedar planking on oak frames, and her overall length was 69 feet, her beam 17 feet and her draft seven feet six inches. She was 70 tons gross weight and 47 tons net. She was built for Roland West, a Hollywood film director, and called *Joyita* (Spanish for "Little Jewel") after his wife, whose name was Jewel. Because of the need for a successful film man to look successful, she was luxuriously equipped with the latest navigational aids, including an automatic pilot, twin diesels, a huge deep freeze, great iceboxes and tanks for 2500 gallons of water and 3000 gallons of fuel. She was a superb sea boat for cruising and was equipped with swivel chairs for big-game fishing.

A great deal of money had been spent to make her a luxury ship, but her whole career was to be one of ill-luck, though that wasn't obvious when West took possession of the vessel. Yet it started early because West's attachment to his wife and to the vessel does not seem to have lasted long. He became involved with Thelma Todd, a blonde film vamp, who died mysteriously aboard *Joyita,* and he sold the vessel to a man called Milton E. Bacon. Thelma Todd's perfume bottles and toilet accessories were still aboard her. When the United States entered the war in 1941, *Joyita* was requisitioned by the U.S. Navy and taken to Pearl Harbor, where she was used for patrolling.

While with the navy her ill-luck struck again and she was run aground off the Hawaiian Islands in 1943. Most of the lower part of her hull had to be replanked. In 1946 she was bought by the firm of Louis Brothers, who converted her into a fishing vessel. They installed two 225-horsepower marine diesels, stripped her interior for conversion to refrigerated holds for fishing and built into her 640 cubic feet of cork in five-inch-thick slabs to act as insulation.

In 1953 she was acquired by Dr. Ellen Katherine Luomala, an American of Scandinavian extraction lecturing in anthropology at the University of Hawaii in Honolulu. A fortnight later she was chartered to Lieutenant-Commander Miller, who was then trading under the name of Phoenix Island Fisheries. Miller always referred to Dr. Luomala as his fiancée and his friends understood that they intended to marry when his divorce was finalized.

Miller, who was 39, was born in Cardiff into a seafaring family. He was a tough, brave man, always optimistic, somewhat feckless in his way of life and a heavy drinker ashore, though he never touched a drop afloat. According to a member of the Suva Yacht Club who knew him, Miller was just the sort who might have commanded one of the Tangier smuggling ships, such as operated on a large scale just after World War II.

He had piercing blue eyes, dark, slightly receding curly hair and a grin that showed the gap in his mouth when he forgot to wear his false teeth as he was inclined to do—especially in his last days when he was down on his luck. He was a little indifferent about his appearance, slight, wiry, virile, sharp-witted, boastful at times, but proud, and while popular with men, he was also attractive to women, though his attitude toward them was somewhat callous because he preferred bars and the company of men. He had joined the British merchant navy at 14 and served with it until the outbreak of war, when he had joined the Royal Navy Volunteer Reserve. He had been demobilized with the rank of lieutenant-commander.

War service in the Pacific gave him a taste for its free and easy islands and he worked in several vessels, always looking for something better and occasionally striking out on his own. He was recognized as a fine seaman, if a little too strict with his crews, but luck never appeared to favor him and instead of improving his position he seemed instead to go rapidly downhill.

As he was a British subject, under United States law his charter of *Joyita* had to be approved by the United States Maritime Administration in Washington. There were no snags and approval was granted for operation in the commercial fisheries at Canton Island and for the transport of frozen fish to Hawaii. That October Miller started fishing, making three round trips between Canton and the Christmas Islands and Honolulu. But the market wasn't right and, unable to sell his fish at a good enough price, he said good-bye to Dr. Luomala in April 1954 and headed for Samoa, once German territory but since 1929 administered partly by New Zealand and partly by the United States. On July 18 he arrived in Apia, Western Samoa, with his holds full of frozen fish, only to find the refrigeration had broken down and much of the fish had gone bad. He tried to get rid of it in Pago Pago, American Samoa, with no more success, and on March 23, 1955, he returned to Apia, where he had to let the catch go for what he could get. He had left Pago Pago heavily in debt and without paying his harbor dues, and because of this some of his ship's papers were retained by the American authorities.

By the end of this trip, his fourth, Miller had to face the fact that he was in serious financial trouble. He asked Dr. Luomala what he should do about *Joyita*, which was lying in Apia harbor. She advised him to sell her, settle his debts and return to Honolulu. He tried the Fiji government but his asking price of $70,000 was too high and, in any case, selling was impossible because no one was willing to buy the ship with her papers incomplete, something he probably forgot to mention to Dr. Luomala.

By this time *Joyita* had been converted many times and, through lack of resources, was beginning to look neglected, with an unreliable radio and engines that were far from good enough for the great expanses of the Pacific. But then Miller met a young New Zealander, R. D. Pearless, who had recently been appointed District Officer to the Tokelau Islands, 270 miles north of Samoa. Pearless was a keen and enthusiastic officer anxious to establish better communications between Western Samoa and the islands under his care. Through him, the Administrator of the Tokelaus granted Miller permission on March 25, 1955, to make a fishing trip in the vicinity of the islands. Pearless went with him, spending a whole month in *Joyita*.

Pearless was only 29, tall, slim, fair-haired, full of plans and ideas, energetic, keen, sometimes difficult to get on with because he was tactless, even supercilious and very much aware of being the white man in the tropics. One of his habits in the Tokelaus seems to have been to have a small boy follow him everywhere carrying a cup of tea in one hand and a saucer in the other. But he was accepted as well-meaning and anxious to help the islanders. Because there were few regular transport services among the Pacific islands, either by sea or by air, except between the capitals of neighboring groups, and because money was not always available for a special charter, it sometimes took months to obtain a passage. The Tokelaus were especially inaccessible, and it seemed to Pearless that *Joyita* could solve his difficulties while, to Miller, it seemed that Pearless could solve his. On their return, they went to see the acting Administrator of the Tokelaus to discuss the use of *Joyita* on a yearly basis. But once again the missing ship's papers proved an obstacle. The government was unwilling to proceed without them and the proposal had to be dropped. By this time Miller's circumstances were becoming desperate and he could not finance any more fishing trips. *Joyita* lay in Apia harbor for over five months, deteriorating all the time.

Miller was now almost on his beam ends, and Commander Peter Plowman, a tall, graying ex-naval man who was the Fijian government's expert on maritime matters, had to give him a meal once when he admitted having had nothing to eat for two days. On another occasion he lived for three days on a small loaf, alone on board *Joyita* because he could not afford to pay his crew. He managed to hide his plight from Dr. Luomala, however, and even managed to avoid selling his last decent suit, though he often depended on the charity of friends and on odd jobs. He once painted the tables and chairs of the Returned Serviceman's Association Club for a few shillings to tide him over. Men who had lent him money had no doubt that if things went right Miller would pay them back, but they were beginning to grow worried and Miller himself was anxious and low in spirits.

Meanwhile the situation in the Tokelaus was deteriorating. The islanders were in need of medical supplies and foodstuffs such as flour and sugar, and they also had 70 tons of copra, the islands' only export, waiting for shipping. Already an attempt to take supplies and lift the copra by flying boat had failed when the aircraft's hull had been ripped open by coral as it taxied in the lagoon.

Pearless was determined to do something for the islands and he approached Kurt von Reiche, the general manager of E. A.

Coxon and Co., a copra shipping firm, to find out whether his company would be willing to charter *Joyita* for one or two trips to the Tokelaus. After much negotiation Coxon finally arranged a charter. Miller couldn't believe his luck. He had just about given up hope when he heard of the charter, and Pearless told a friend that he seemed stunned. He said, "This is the happiest day of my life."

There was one stipulation: Coxon had two officials who had to go to the Tokelaus so that *Joyita* would have to carry passengers, something she was not licensed to do under the American regulations under which she operated. Miller decided there were ways around this problem, but his ship was not insured (perhaps because he knew she couldn't sink because of the cork in her hold) and he had no crew. Nevertheless, two faithful Gilbert Islanders he had employed, Tekoka, the bosun, and Tanini, the engineer, gave up the jobs they had found in Apia and rejoined him. They weren't particularly enamored of the elderly *Joyita* but they were devoted to Miller, who planned to go to Honolulu eventually and take Tanini with him. As mate, he managed by a stroke of luck to sign a powerful American, Chuck Simpson, or Captain Jah, as he liked to be known, a man with a huge chest and powerful arms like a gorilla's. It was fortunate that he was available because, under U.S. regulations, *Joyita* had to have an American national aboard as an officer. Simpson, a part-Indian, had married a Samoan girl and was working in an Apia garage. Miller scraped up a few more men, but it wasn't easy because *Joyita*'s reputation was well known. At the last moment he obtained a young part-Samoan, Henry McCarthy, who, though he had no real training as an engineer, was very capable, having picked up the trade from the American Marines during the war. His mother knew *Joyita* well enough to be unhappy about his signing on as a member of her crew, but he reassured her and spent a lot of time aboard cleaning up the neglected engines. When they were due to leave, he told her that everything was in fine shape.

A substantial amount of cargo was put aboard and *Joyita* was joined by nine passengers, including Pearless, Dr. A. D. Parsons, a doctor from Apia Hospital, Mr. J. Hodgkinson, a dispenser from the hospital, and seven Tokelau Islanders stranded in Samoa, including a woman and two children. There were 16 crew, and Pearless and the two officials from Coxon, J. Wallwork and G. K. Williams, signed on as supercargo because *Joyita* was not licensed to carry passengers. Miller decided to ignore the regulations with regard to the others.

Aged 41, Dr. Parsons was an Irishman from Athlone, dark-haired, ruddy-faced and good-looking, but he had been badly wounded during the war and walked with a limp, and the head injuries he had received affected his temper so that, on occasion, after a few drinks he could become violent and dangerous. Hodgkinson was a quiet man of 49, married with three children. Wallwork and Williams were making the trip to do the buying and trading of the copra, and Williams, who carried with him £1000 in Samoan currency—£950 in notes, £50 in coinage —to pay for the 70 tons of copra, half of which was to be shipped this trip and half on the next, was 61 and married, the retired manager of the Yorkshire Insurance Company in New Zealand. They were heading for the port of Fakaofo, 270 nautical miles away. Yet unexpectedly 2640 gallons of diesel were put aboard on credit, giving the vessel a range of 3000 miles. There was also enough food and water for a much longer voyage.

Everybody was aboard by 11:00 A.M. on October 2 and *Joyita* prepared to leave Apia soon after noon. Her engines were started and she began to move toward the open sea. Commander Plowman had a house which overlooked the harbor and as the vessel neared the harbor entrance he saw a puff of smoke burst from the ship's side and *Joyita*'s forward movement stopped. As she began to drift toward the reef an anchor was hurriedly thrown out and she came to a halt. Concerned, Plowman telephoned the acting Commissioner for the islands to ask if the government officials shouldn't be taken off, and, worried, they discussed the matter for a while before finally deciding to leave it until the next day.

Joyita lay there all day and was still there when darkness fell. Rather than endure the discomfort aboard, some of the passengers returned ashore, bringing the information that the port engine was not working properly and would take several hours to repair. The Europeans made for their clubs while *Joyita*'s engineers struggled with the problem. Von Reiche went on board and told Miller, "If you are having any trouble do not put to sea until it is right." Miller reassured him that they had had the trouble before, that it was a simple matter and that they could easily fix it, even on the way. He was doubtless worried that if he didn't leave quickly, word might reach the American authorities and they might prevent him from making this all-important trip which was to recoup his fortunes.

Toward midnight the passengers reembarked to find that the starboard engine was running but that the port engine clutch was still not repaired and was partially disconnected. *Joyita* finally left at 5:00 A.M. on Monday, October 3, heading north, with the crew exhausted and the passengers already angry and on edge. When Plowman woke, prepared to insist on the removal of the government officials, there was no sign of the ship.

With full holds and 25 people on board with all their luggage, to say nothing of the food needed to feed them, it must have been uncomfortably crowded. The author served for two years in vessels of the size of *Joyita* with crews of 12 or 13 and even that could be uncomfortable, especially for sleeping.

The people waiting in Fakaofo, in the Tokelaus, were happy to learn that the ship that was to bring them their badly needed supplies and take away the copra they had collected was on her way. But when the days passed and she didn't appear, they began to wonder what had happened. As more days passed, they grew concerned and, when she was finally reported overdue, the Samoan authorities informed the Royal New Zealand Air Force at Laucala Bay, Fiji. A Sunderland left on October 6, three days after *Joyita* had left Samoa, and flew to Apia and from Apia to Fakaofo. On the return trip it flew a course 20 miles to the west of its northbound trip, but despite good visibility and radar sweeping 10 miles on either side of the aircraft, nothing was found. From then on until the search was abandoned on October 12, nearly 100,000 square miles of sea were covered. Not a trace of *Joyita* was found. All ships and radio stations within the area where she might conceivably have drifted if she had been disabled were informed, but no message had been received. Finally the search was called off.

To those people who knew Miller, it seemed that his ill luck and his stupidity in going to sea with faulty engines had done for him at last.

Joyita had been completely written off when, at 0654 (Fiji time), November 10, 37 days after she had left Samoa, a message was received at Suva, Fiji, from the Gilbert and Ellice Islands' supply vessel *Tuvalu:*

> *Joyita* found waterlogged in position 14 degrees 42' south 179 degrees 45' east by dead reckoning. Boat sent across but nobody found on board but possibility in flooded compartments. Port side superstructure including funnel blown or washed away. Canvas awning rigged apparently subsequent to accident. No log book or message found.

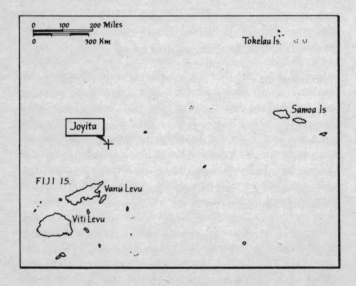

Captain Gerald Douglas, master of *Tuvalu,* had spotted the
derelict about three miles east of his course and 167 miles north
of Vanua Levu, Fiji, some 450 miles west-southwest of Apia. As
he had closed her he had seen the name on the bow was that of
Joyita. She was listing 55 degrees to port, with her port rail well
awash and only about three feet of her starboard side above water. Though her hull was undamaged, the monkey island and
steering position built above her deckhouse had been smashed
away and a canvas awning rigged. Her course appeared to have
been in almost the opposite direction from the Tokelaus. Though
she should have headed north for 270 miles, she was found over
600 miles from her destination in a west-southwesterly direction.
A moderate easterly swell was slapping against her starboard side,
causing her to roll sluggishly, and Douglas ordered away his boat
under his chief officer. There wasn't a single soul aboard her or
any indication of what might have happened to them.

The Fijian tender *Degei* was sent to tow the derelict in, and
when Captain Robert James checked, he found no bodies, no
lifeboats or rafts, no food in the galley or in any other compartment above water, no log and no sextant. Towed to Vanua Levu,
a port with limited facilities, she was put on the nearest beach
and the pumping out started.

About her there was an unpleasant odor of decay which came
from the soggy wreckage littering the cabin, but the experts
agreed that she was still fundamentally sound. Barnacles above
her normal waterline indicated that she had been listing steeply
for some time, but they felt that the hull damage had been caused
by the breaking waves smashing against her. Whatever had happened to her seemed to have happened at night because the clock
had stopped at 10:25 and the switches for the cabin and the masthead navigation lights were in the "on" position. There were indications of an engine failure, one engine being covered with
mattresses and the other showing that repairs had been attempted.
The deck awning at first seemed to indicate that passengers had
been sleeping on deck because of the tropical heat or the lack of
bunks. Almost every movable object useful at sea had disap-

peared and there was no sign of the lifesaving equipment, a dinghy powered by an outboard, a 16-man life raft and two 10-man rafts.

In the meantime, on the receipt of Douglas's message, a new air search had been made in the area where the derelict had been found and near the Ringgold Islands to the southeast. Not a trace was found of the three carley rafts *Joyita* was known to have had aboard when she left Apia.

Newspapermen, scenting a story, seized on the event joyfully, invoking as always the ghost of *Mary Celeste*. The men searching her found no message, as they had half-expected. Some hasty words could easily have been scribbled on the deckhouse or on the canvas awning that had been rigged, but there was nothing, and they wondered if a huge wave had smashed down on her, to crush the superstructure and carry everyone on board away. But, even as they thought about it, they rejected the idea. Out of 25 people on board, someone would surely have been below and survived to cling to the ship, waterlogged as she was. There had been a case in March 1907 of people living aboard a waterlogged ship for weeks. The American schooner *Everest Webster* had been found by the four-master *Quevilly,* a wreck, apparently abandoned, almost totally submerged and with only her stern above water. But when boarded, a group of "corpses" found in the bunks proved to be still just alive and starving. She had been wrecked early in the year and had floated half-submerged ever since, known about and a menace to shipping and with a coast guard vessel on the watch to sink her.

And why abandon *Joyita* anyway? This question came up again and again as theories were advanced. Miller knew his ship was almost unsinkable and must have known it was safer to stay with her and hope to be spotted by an air search. Indeed, he had often said that in the event of trouble he would stay with the ship and his friends firmly believed he would never have left her of his own accord. The fact that she was unsinkable was obvious because she had been found half under water in a condition which would have taken any other vessel to the bottom.

As the pumping out proceeded there was an alarm as a Fijian

workman examining the hold shrieked out that he had found a body. But it appeared that, groping in the dark, he had put his hand on a child's rubber ball and, finding it soft and cold to the touch, had leaped to the conclusion that it was a corpse.

Since the vessel was not holed it was hard at first to account for the flooding. An auxiliary pump had been rigged up in the engine room, lashed to a length of timber laid across the two main engines. But it had not been connected, though the presence of four mattresses in the engine room seemed to suggest there had certainly been trouble there. The port main engine clutch was not working and was partially disconnected. The radio had been tuned in on a wavelength of 2182 kilocycles, which was the distress wavelength for small ships and seemed to indicate that someone had made an unsuccessful attempt to get off an SOS.

What had caused *Joyita* to become waterlogged was the first question to be settled. Captain E. L. James, assistant harbormaster at Suva, and Rob Wright, the Public Relations Office photographer, were sitting aboard *Joyita* chatting when they heard the sound of water entering the vessel. Puzzled and knowing she was supposed to be sound, they started searching. In the engine room they found water pouring through a one-inch pipe on the port bilge, well below the floor boards. The galvanized pipe had been threaded into a brass T-piece in the salt-water cooling system, but, since ferrous and nonferrous metals joined together could set up an electrolytic action in tropical sea water, the pipe had corroded completely through and a section had actually fallen off. Wright took photographs at once.

At least they had identified the cause of the waterlogging. When the vessel had been found the presence of the four mattresses in the engine room had been puzzling, but now they realized they had probably been stuffed in there to block the leak when it was located. Because the pipe was situated nine inches below the engine room floor and because of the noise of the diesel, it had obviously not been traced. The first thing *Joyita*'s people would have known about it would have been when the water rose above the floor boards, by which time it would have been difficult to locate the source of the trouble.

It was also soon discovered why no SOS had been sent. There was a break in the aerial lead 18 inches above the transmitter. Certainly somebody had tried to transmit a distress signal, but the break, which would not have been noticed because it had been covered with green paint when the deckhouse had been painted, would have reduced the range to about two miles.

Why had *Joyita* not been spotted by the searching aircraft? Men involved in search and rescue by air and sea knew only too well how difficult it was to spot something small, especially something barely afloat. Also, being a wooden-hulled vessel, *Joyita* did not give a clear response on radar like a metal ship. It was noticed as she was being towed by the tender *Degei* that the steel vessel showed up clearly on the radar screen at 20 miles but *Joyita* projected no image even when the aircraft was close.

But there was still no explanation for the disappearance of the 25 people on board. It seemed clear that *Joyita* had been abandoned early in the trip. The current in that area sets in a west-southwesterly direction, so that, had she been abandoned toward the end of the trip, she would have been found farther to the north.

Commander Plowman, as the Samoan government's expert, had been sent to look at *Joyita,* and, searching the derelict vessel, he immediately made some interesting discoveries that led him to conclude that no attempt had been made to throw out a sea anchor. Though the ship probably did not carry a ready-made one, one could easily have been fashioned from ropes and a length of timber, of which there were plenty on deck. He also decided that the awning, which had been fixed aft of the bridge, was presumably a protection against the sun or for catching rain water, and considered that it had been rigged after the accident which had damaged the port side superstructure because it was lashed to a piece of broken stanchion. Since the knots were half-hitches, he decided it had probably been rigged by someone who was a sailor.

More ominous, he found in the scuppers a waterlogged doctor's bag containing a corroded stethoscope, a scalpel, some

needles and catgut and four lengths of bloodstained bandage. He
concluded that someone had been hurt and decided that it must
have been Miller, because only if Miller had been out of action
would the rest have taken to the life rafts, something Miller,
knowing the vessel to be virtually unsinkable, would never have
permitted.

By this time the news had flown around the world and strange
rumors were cropping up in the Suva bars, to which the government added by making contradictory statements. The first of
these suggested that *Joyita* had been run down by a ship and that
the first examination of the derelict had revealed damage to the
port side. The following morning, however, the government announced that *Joyita* had now been pumped out and was floating
upright. Inspection had revealed that there was no holing and no
sign of fire, and there were no bodies aboard. However, all the
cargo had disappeared! Seven cases of aluminum strip—used for
rat guards on coconut trees—had gone from the midships hold,
and 44 pounds of flour, fifteen 70-pound bags of sugar, eleven
56-pound bags of rice and 460 empty copra sacks were missing
from the after hold. In addition, the £1000 carried by Williams
was also missing.

Towed to Suva and hauled up the slip, *Joyita* immediately
started talk in the Fiji Club, the Defence Club and the Royal
Suva Yacht Club. Why had no one found time before abandoning *Joyita* to scrawl a message somewhere? Why abandon her
anyway? Had there been foul play? What had happened to the
cargo? And had the awning really been rigged after the accident
that had damaged the superstructure and, if so, by whom?

Three theories were going the rounds. A sea quake had thrown
the ship on her beam ends. This was known to have happened six
months before to the Tongan vessel *Fifofua* when 40 passengers
had been flung into the sea. However, except for one child, who
was drowned, they had all managed to scramble back on board.
Since charred planks and a door had been found, a second theory
was that there had been an explosion on board, but this was discounted when it was discovered they were the result of a small

fire on board before the last voyage. A third theory, springing
from the damage to the wheelhouse, was that *Joyita* had been
rammed and looted.

A waterspout was also suggested when it was believed that
John Williams VI, a missionary schooner whose crew had seen
seven waterspouts about the time *Joyita* had been abandoned,
had been in the area, but this was proved to be untrue. Sea monsters, tidal waves and smugglers all had their period of popularity,
as did some even stranger explanations. For example, it was suggested that she had been abandoned because the people on
board, deprived somehow of Miller's advice, had decided that
Joyita was about to sink; while Leslie Hobbs, of *Life* magazine,
argued that they had tried to return to the ship but that in the
dark they had failed to locate her and had lost their lives trying to
reach land. Hobbs also raised the possibility of pirates and mentioned that the searching New Zealand air force aircraft had photographed vessels of a Japanese fishing fleet across the course
Joyita must have taken.

Everybody in Fiji had his own theory. The *Fiji Times and
Herald* seized on the story like a dog on a bone. "The locating of
Joyita deepens the mystery that has shrouded the circumstances
of her disappearance," the paper announced, and at once set
about castigating the government. "There has been a great deal
of official secrecy. . . . What is it all about? Wherever our reporters have gone in an attempt to present a true picture to the
people of an event of world-shaking interest they have been
frustrated by officialdom."

Overseas newspapers, fed by agencies, had their own views.
The *News Chronicle* suggested that the damage to the wheelhouse
indicated that *Joyita* was the victim of a hit-and-run raid. It was
obvious that the newspapermen weren't being too careful about
their theories because they were getting their facts wrong. Knowing nothing of the cork lining for the freezers, they reported that
Joyita had been kept afloat only by the empty oil drums she had
on board. The reports also indicated that the cabin was badly
charred after a heavy fire and that the survivors must have es-

caped on a raft of timbers and oil drums, because there was no trace of the drums and 2000 feet of timber she was supposed to have carried.

The *Daily Telegraph* suggested that unsurrendered Japanese soldiers or sailors were responsible. There were still many of them in the Pacific in 1955, some perhaps imagining in their lonely island strongholds that the war had not yet ended. One rumor even credited them with a submarine, kept concealed in a camouflaged pen in a bay on their island base, and it was suggested that other vessels which had disappeared in the Pacific since the war might have met the same end.

Another suggestion, springing from Leslie Hobbs's hint in *Life* that one of the Japanese fishing vessels might have had piratically minded men aboard, came from the *Fiji Times and Herald*. This time the paper went so far as to offer a minor massacre. "ALL ABOARD JOYITA MURDERED," it announced, even hinting that this was the official line of thought. *Joyita,* it said, had probably run through the Japanese fishing fleet. Forty-eight boats had been seen in the vicinity about October 4 and it was suggested that *Joyita* had spotted something the Japanese had not wanted them to see.

"There was a young District Officer on board who possibly protested, . . ." the paper went on, and added that perhaps resentful Japanese had boarded *Joyita,* murdered or taken prisoner the passengers and crew, attempted to blow up the ship and opened the seacock. The report went on to point out that a steering chain on one side of *Joyita* was broken, and that this was also thought to have been done by the Japanese and that the damage was not the result of a collision but a deliberate attempt to wreck the vessel.

The report continued:

The *Fiji Times and Herald* informant stated that the *Koyo Maru,* mother ship of the Japanese fleet, was one of the most modern ships in the world and was fitted with the most modern radar afloat. It is recalled that on her last visit to Suva,

when she called ostensibly to arrange salvage, shore leave
was not granted the crew, who, on the visit before, had
swarmed all over Suva.

The Japanese were certainly unpopular and there was a great
deal of resentment that, after all the atrocities that had been
credited to them during the war, they should be allowed to send
their fishing fleet to operate in local waters, but the *Fiji Times and
Herald* was being a little too vigorous at stirring up trouble. The
government was furious at the suggestion that the boarding and
murder theory was the official line of thought. Mr. A. F. H. Stoddart, CMG, the Colonial Secretary of Fiji, even took the unusual
step of appearing on Suva Radio to deny the story emphatically.
He said it was rash to accuse the Japanese and outrageous to
suggest that the view was an official one. He pointed out that the
government had been most careful *not* to express opinions about
the disaster and did not intend to do so until a full examination
had been made of the vessel. No seacock had been found open
and the searching planes had not photographed the Japanese
vessels. He added that the evidence, in fact, seemed to suggest
natural causes and that *Joyita* appeared to have been overwhelmed by freak weather when perhaps her engines were
stopped. He pointed out, however, that this view too was pure
speculation.

But the official announcement didn't stop the speculations. The
newspapermen had the bit between their teeth and were wringing
every possible scrap of drama from the story. All around the
world there were rumors of gun-running and the version of massacre by the Japanese reappeared dramatically when it was reported in the *Fiji Guardian* that Japanese knives had been found
aboard *Joyita*. This time the story was not officially denied because it was true. Two men cleaning up *Joyita* had found knives
stamped "Made in Japan" and had handed them over to the police, who were testing them for bloodstains. The story died a natural death when the tests proved negative and the police admitted
that the knives were old and broken and might well have been
aboard *Joyita* when she was used for fishing.

Nevertheless, it was known that Williams had been carrying £1000 in cash and it was even suggested that the poverty-stricken Miller had a similar sum with him. Further stories indicated that two compasses and two radios were missing from Miller's threadbare craft and that, when the vessel was found, the refrigerators had been full of meat which should have been taken by the survivors as they abandoned her. Even the saloon furniture was said to be missing, but, with the number of people aboard and the state of Miller's craft, there could hardly have been much. Had the crew planned to scuttle her? With all the other people on board this idea didn't hold water. Even Mary Pickford, the film star, was dragged into the story when it was suggested that *Joyita* had belonged to her.

Then drug-running as a reason for the disaster came up when a letter arrived describing how the writer had been approached by a 50-year-old Chinese who told him there had been opium aboard *Joyita*. The letter, which had been posted in Sandringham, Victoria, bore an illegible signature, but the envelope also contained a scrap of paper containing the words, purported to have been written by the unknown Chinese, "Me no Joyita got Big Opium in Bag of Pototo. I am Chinaman she got Big Lot."

A Mrs. E. Cromer, of 42 Toongabbie Road, Toongabbie West, New South Wales, wrote that she had been in contact with the planet Mars and, knowing the real story, hoped the Japanese would not be blamed. According to Mrs. Cromer, a drum of gunpowder had exploded and the fumes had been so choking everybody had leaped overboard. It sounded almost as if she'd been reading about *Mary Celeste*. The drum of gunpowder, she went on, together with a drum of glycerin stacked on top of it, had been blown through the side of the ship which had afterwards been boarded by fishermen. This story was said to have come to Mrs. Cromer from one of the dead.

Stoddart, the Fijian Colonial Secretary, even received a letter from a man who called himself "Bishop Bridge" and claimed to be his "creator and savior." Bridge apparently lived in Westor, New South Wales, and suggested that if Stoddart cared to investigate, he would find everybody from *Joyita* in the Solomon

Islands, "happily cured of all their cancers and TB," having been "flashed" there by the Bishop with a "miracle."

These were not all. Miller's dubious dealings were remembered and it was said that he had had no intention of going to the Tokelaus at all. Why take on all that water, fuel and provisions for such a short trip? It was suggested that, having reached the end of his credit in Samoa, he intended to sail to Honolulu. It was remembered that long before the disaster, Mrs. Ismay Beatrice Matilda Miller, his estranged wife, had begun divorce proceedings against him in his hometown of Cardiff and it was suggested that he had disappeared to avoid paying alimony. But how could he have gone to Honolulu against the protests of 24 other people, all of whom urgently wanted to go to the Tokelaus? *Joyita*'s position when she was found did not seem to indicate that she had been heading north-northeast toward the Hawaiian group. Nevertheless, it was felt that Miller was clever enough to try it and had been desperate enough to need to. The government was sufficiently rattled to issue a statement to the effect that "no good purpose could be served by uninformed speculation as to the cause of the *Joyita* disaster until the date of the commission of enquiry."

The court of inquiry sat at Apia, Western Samoa, on February 3, 1956, under the chairmanship of Judge Carsack. Commander Plowman described *Joyita*'s false start and Kurt von Reiche, general manager of Coxon, described going aboard her on the Sunday she was due to sail and telling Miller not to sail if he was having engine trouble. It seemed likely, he said, since the port main engine had its clutch partially disconnected, that Miller had left on one engine. He said Miller had told him that two more trips like the one he was about to make would clear up all his debts and enable him to buy fuel to go to Honolulu.

There had been some alarm, he went on, when it was noticed that one of the cases being loaded aboard the vessel was seen to be marked "Highly Inflammable. 75 degrees flash," but it was shown on the manifest that the case contained not explosives but rat traps, and that was what it had proved to be. Von Reiche also revealed that Miller had decided not to take the vessel's powered dinghy and had left it with him, claiming it would not be required for the trip, and said that while von Reiche was drinking with Dr. Parsons, Miller had also been offered a drink but had refused, saying he never drank while he was afloat.

Superintendent Bentham of Apia Radio said that when it was announced that *Joyita* was to sail, he sent a message to Miller, since no request had been received about working with her, advising him that the radio station was available to work with the vessel at 10:00 A.M. and 4:00 P.M. daily. He had no authority to insist on this or to give approval to *Joyita*'s radio equipment, but he suggested that it would be wise for *Joyita* to be in contact and that Miller should test his radio before leaving.

When no word was received, Bentham went to the wharf, where he asked Pearless to get Miller to make the test and to find out if the times he had suggested for working with the ship were satisfactory. Again no word came back. He waited at the radio station until seven on the Sunday evening, certain that some steps would be taken to check *Joyita*'s equipment. Still nothing was heard, but about noon on Monday, after *Joyita* had sailed, von Reiche called at the radio station and advised Bentham that *Joyita*'s call sign was WNIM and that Miller would work at the times suggested. From then onward and throughout the whole period of the search, Apia Radio listened out and called *Joyita* constantly, but without response.

It was then revealed that *Joyita* was not licensed to carry passengers and that both Pearless and Miller must have known of it. Miller certainly had and Pearless must have learned of it because it was in the terms of the agreement he had made. And since the two Coxon officials were also on board as supercargo, it was just possible that Coxon and Co. were also aware of it. The fares were £13 for administration officials, £6.10.0. for islanders and £3.5.0. for one of the children who was under three years.

By this time Miller, who had always been a popular and good sailor, could be seen as a man desperate enough to cheat to gain what must have been his last chance. Not only had he told von Reiche the engine could easily be fixed, when he must have known that it couldn't, and refused to have his radio tested because he probably knew it needed repairing, but he had also lied to the Apia harbormaster about having a license to carry 25 passengers issued by the Bureau of Shipping, Honolulu. Doubtless he feared that the charter, his last chance, might be called off if there was any delay. Evidence was also given that Miller was down on his luck and anxious, but it was admitted that he was a good sailor and that in an emergency would always have stuck to his boat.

Mr. T. R. Smith, the acting High Commissioner, had kept a careful log of events concerning the search for *Joyita,* and the court learned from him that Miller had been under pressure from Dr. Luomala to return to Honolulu and that he might indeed have gone there. He also revealed that Miller had insisted on 60 drums of fuel being placed aboard instead of the 30 which were ample to take him to the Tokelaus, and the diary also showed that the wife of Chuck Simpson, the mate—and therefore surely Simpson himself—knew the Morse radio was out of order and that there was no other radio aboard. Both her husband and Miller had said, however, that they would be back from the Tokelaus on the following Friday and, since Miller had also left behind a suit and several shirts, it did not seem that he intended to disappear.

Smith also revealed that von Reiche had given him a letter from Miller to be posted to Dr. Luomala in Honolulu but that, under the circumstances, he had felt justified in opening it. It showed that Miller had had no intention of going to Honolulu until after the Tokelaus trip, which he hoped would make enough money to pay his debts. He seemed keen to help the Tokelaus and was hoping for future charters, and the sheer difficulty of persuading everybody on board (all of whom urgently wanted to go to the Tokelaus) of the need to go to Honolulu instead is enough to preclude the possibility that Miller tried to go to Honolulu.

Smith also pointed out in his evidence that it was most unlikely that Dr. Parsons and Hodgkinson, the dispenser, from the hospital, would have agreed to any change of plan because they were on a mercy trip to perform an emergency operation. The evidence, in fact, finally seemed to indicate that there was little likelihood that Miller had ever wished to go to Honolulu.

Twenty-eight witnesses, many of them expert, explained the cause of the flooding, the breakdown of the engines and pumps, the failure of the electric light and the uselessness of the radio. The court's report, issued on February 22, 1956, rejected all the fantastic theories that had been offered and found that the cause of the disaster was the flooding of *Joyita* through the corrosion and breaking of the one-inch galvanized pipe forming part of the cooling system of the port auxiliary engine, and that the most important contribution to the disaster was the failure of the bilge pumps, probably by the blockage of the intake pipes by rubbish and the absence of strainers. With no watertight bulkheads, the water had found its way through the ship, and as the engine had stopped and she had lost steerage way, she had become unmanageable and fallen broadside onto the seas, so that her master and crew were powerless. Her natural buoyancy, together with empty drums wedged into the holds, were enough to keep her afloat.

The commissioners, however, had no idea at all what had happened to the people on board. Though willing to make a decision about what had happened to the ship, they felt "quite unable to do so with regard to the ship's personnel." They felt the launch was abandoned quite deliberately but were unable to explain why, as the awning lashed to the damaged superstructure seemed to suggest that there were survivors on board after the hull had filled. After the disaster the failure of the radio and the absence of a properly equipped lifeboat contributed to the casualties. The court suggested there might be living survivors on some of the hundreds of islands that dotted the area where *Joyita* was found, though they regarded the possibility as remote.

They felt also that Miller carried a heavy responsibility for his failure to provide a working radio, a lifeboat or other lifesaving equipment. He was also negligent in not making a trial run before leaving Apia, as that would have shown up some of the defects resulting from *Joyita*'s having lain idle for five months.

The inquiry was considered clear and honest, but in fact not all the evidence had been brought to light. Von Reiche had not told the court how on the Sunday evening he had gone aboard *Joyita* a second time, again asking Miller if it was safe to sail. He had received the reply, "Don't worry. It'll be all right. Besides, if anything did happen to this boat a man would be a fool to leave her because she's unsinkable." It was clearly because of this knowledge that Miller had felt justified in taking her to sea in such bad condition.

What was probably more important, no evidence was given of the discovery in the scuppers of the doctor's scalpel, needles, cat-gut, stethoscope and four lengths of bloodstained bandages. Considering his discoveries interesting, Commander Plowman had expected to be called at the inquiry, but on the last evening, the Attorney General, William Watson, had told him his evidence had already been covered. Plowman did not consider that it had. It was his view that the court was trying to cover for the government, which should never have allowed *Joyita* to sail and was well aware of it.

So what had happened?

The Pacific was always throwing up mysteries. In 1940 the ketch *Wing On* was found with three sails rigged and the tiller lashed, canted on a reef near Vanua Levu. She was apparently deserted, but the men who boarded her from a mission vessel found a starving white woman in the cabin, with only her head above the water that almost filled it. When she was dragged out they realized she had been trying to hold up the body of another woman, and also on board was the decomposing body of a man. The woman was the sole survivor of two young married couples who had tried to sail from San Francisco for the Marquesas. They had not taken enough precautions, their boat was leaky, had faulty pumps and a useless chronometer, and, becoming lost, they had run out of food and three of them had starved to death.

In 1956 the 130-ton steel twin-screw motor vessel *Melanesian* left Sikiana Atoll with 64 people on board on an inter-island service trip, and after reporting herself near Sikerane was never seen again. Some time later a piece of timber was found, together with the body of the bosun and a buoyancy tank, both of which had been subjected to tremendous pressure. The court of inquiry decided that she had not been sunk by a mine, as was suggested, but had been smashed by some unknown force which could not be identified. Explosions had been heard and local people felt she had been sunk by an explosion in the battery room, but at that time strange things were taking place in the Pacific and two years before, not far west of where *Melanesian* disappeared and not far north of where disaster overtook *Joyita,* the Japanese fishing boat *Lucky Dragon* had found herself enveloped by the fallout from an A-bomb test on Bikini Atoll.

In 1958 the yacht *Annette,* which had left Apia on a trip to Suva, was found sunk on Dibble's Reef about 10 miles north of Vanua Levu. One of the men who went to investigate her was Rob Wright, the official photographer from Suva who had photographed *Joyita. Annette*'s rigging was intact but she had been holed by coral. Everything seemed to be in place, but there was no sign of the married couple who had been aboard her and her dinghy eventually floated ashore 60 miles away. Though the outboard engine was missing, a pair of oars and a plastic water bottle were still inside, a strange thing if the boat had overturned and drowned its occupants.

In 1960 the brand-new 114-ton fishing vessel *Teiko* vanished while on passage from Tonga to Samoa. No radio signal was received and none of the 22 crew was ever seen again.

So what did happen to *Joyita*? Had Miller intended to go to Honolulu (the extra fuel and provisions seemed to require an explanation), and was the letter he had written which fell into the acting High Commissioner's hands merely a cover-up? Had he known the letter would be opened when he failed to turn up? Yet it still didn't make sense that he could have persuaded 24 people to go in a direction they did not wish to travel, unless perhaps the course had been changed after dark when they were asleep.

The one thing any sailor learns very quickly is that it is always foolish to take a chance with the sea. Yet experienced men, through force of circumstances, still do so, and Commander Plowman considered that Miller had sacrificed sound seamanship to get money from the charter—believed to be worth £70 a day.

6. M.V. JOYITA (1955)

He felt Miller would have been stopped from sailing if he had still been in the harbor on the Monday morning. During Sunday night Plowman had awakened to find the wind blowing at about 40 miles an hour and had thought to himself, "Thank God I'm not on board *Joyita*."

Captain S. B. Brown, of the 100-ton auxiliary ketch *Maroro*, who was a friend of Miller's, was convinced Miller would never have left *Joyita*. Brown felt that he had had a row with the huge American mate, that they had come to blows and that during the struggle they both fell overboard, which left the ship without a single experienced officer. He was certain Miller would never have left the vessel alive.

The divorce Miller had been waiting for came through in November 1960, when the Divorce Commissioner in Cardiff granted Mrs. Miller a decree *nisi*. He referred to Miller as "shiftless."

Joyita lay in Walu Bay, Suva, until July 1956, when Dr. Luomala put her up for auction. She was sold for £2425 to David Simpson, a planter from Vanua Levu, and was overhauled and refitted. She went to sea again in 1956, surrounded by false rumors that the U.S. government was about to seize her because Dr. Luomala had transferred her registration from the United States to Britain without permission. Still dogged by ill-luck, on January 8, 1957, while carrying 13 passengers, *Joyita* ran aground in the Koro Sea on Horseshoe Reef, a ring of coral about a mile wide. She was eventually hauled off and towed to Levuka, but by this time the islanders were beginning to believe she was haunted. In October 1958 she was recommissioned and began to trade regularly between Levuka and Suva, but on November 20, 1959, she ran aground again on a reef at Vatuvula, though she managed to float off at high water. Examination at Levuka showed little damage but, while on the interrupted journey to Suva, she began to take in water. The pumps were started, but the water came faster and she had to return to Levuka where it was discovered the engineer had fitted the valve box of the pump the wrong way around, so that instead of pumping water out, they had been pumping it in.

It was the end. She was clearly a bad-luck ship and her owners left her on the beach and let her creditors help themselves to her fittings. She was virtually an abandoned hulk when she was found by Lord Maugham—Robin Maugham, the author, and a nephew of Somerset Maugham, who had written so much about the Pacific. He was immediately intrigued by the story and not only conducted a private investigation but also bought *Joyita*.

He came to the conclusion that the evening after *Joyita* left Apia there was rough weather during which the port engine stopped and the one-inch galvanized pipe broke below the engine room floor. The bilge pumps were unable to handle the inrush of water and, anyway, the suction pipes were choked with cotton waste and other rubbish which had collected below the floor boards. As the ship began to fill, the four mattresses were taken below, either to block the leak when it was finally found or to prevent spray being thrown up by the flywheel from short-circuiting the electrical equipment.

The radio was tuned to send a distress signal, but the break in the aerial lead prevented its being heard, and, as the pitching grew worse and the water continued to rise, an auxiliary pump was roped to a piece of timber laid across the two main engines but there wasn't time to connect it. When the water reached a height of 18 inches above floor level the starboard engine also stopped and the ship was plunged into darkness, lost steerage way, became unmanageable and lay beam-on to the seas.

The ship's superstructure had been built of lightweight timber when she had been converted into a fishing vessel and was not as strong as the rest of the vessel. Maugham decided that at 10:25, when the clock stopped, a huge wave had smashed down on it, carrying away the port side. Water poured into the hold and the ship began to flood.

Some time later, he felt, the canvas awning was rigged from a point just above the upper wheelhouse by one of the crew, using half-hitches, and at some point someone was badly hurt and attended to, presumably by Dr. Parsons—though the dispenser, Hodgkinson, could also have done it—using the medical equipment and bandages which were found. At some stage, too, the three carley rafts were launched and the deck cargo, chronometer and the £1000 in Samoan currency removed and everybody left or was taken off the vessel.

Maugham developed a theory which he laid before Judge Carsack, the chairman of the commission of inquiry, who agreed with it. Maugham thought that when the vessel left the second time after her initial breakdown in the harbor, the starboard engine was running but the port engine clutch had not been repaired and was partially disconnected. Toward noon a stiff breeze got up and the vessel, working on one engine, began to labor. The islanders lay soaked on deck while Simpson and Tanini worked in the engine room, helped occasionally by Miller.

The weather grew worse and when Miller went to the bridge, the engineers were helped by McCarthy, the American-trained part-Samoan. After dark the storm broke and the squalls battered the vessel. Parsons tried to persuade Miller to turn back, but he refused and when Parsons tried to force him, they struggled. As they did so Miller wrenched himself free but staggered and fell from the bridge to the deck, receiving a bad head wound. He was at first thought to be dead, but Hodgkinson fetched Parsons's bag and, finding him still alive, Parsons stitched up the wound with the needle and gut found by Plowman, stanching the blood with the bandages he had found. Working only with a small light, they were watched all the time by Tanini, the engineer and Miller's faithful friend.

By this time the water had risen further and the ship was in darkness. Attempts to transmit a distress signal were made but there was no acknowledgment because, with the break in the aerial lead, the range was no more than two miles. With no one in charge, both Pearless and Parsons were trying to give orders when a freak wave crashed down on the port side of the vessel, smashing the wheelhouse and tearing away the superstructure.

With *Joyita* apparently on the point of sinking, the carley rafts were thrown into the sea and attached by line to the vessel, which was now listing heavily to port. Panic caught the passengers and many of the crew, and they swarmed into the rafts. Some of the

islanders and Europeans remained on board and the faithful Tanini refused to leave Miller.

Then Simpson saw what looked like a reef in the moonlight. A
reef indicated land and, with the wind blowing toward it, it
seemed they ought to be able to reach it. Everybody now got into
the rafts, Williams worrying about the £1000 until he realized
that even if it were lost the Bank of Samoa would replace it—as
in fact it did.

He joined the others in one of the rafts as *Joyita* lurched to
port, diving into the waves. The rafts were cut adrift, but unfortunately no one had remembered that the current set due west so
that the rafts, instead of drawing nearer the land, drew farther
away and they were never seen again.

Miller, who was still on board *Joyita,* guarded by the faithful
Tanini, died some time after the storm abated, and the abandoned
ship with only Tanini on board alive drifted across the empty
sea. Eventually she was sighted by a Japanese fishing boat of
about 80 tons whose crew boarded her. They found Miller's body
but Tanini was still alive, though starved and no longer sane, and
rushed at the nearest Japanese to drive him away from the body.
During the struggle Tanini lost his balance and fell overboard.

Now alone on board *Joyita,* the Japanese, who had come not
as pirates but to help, decided, knowing nothing of the cork lining, that the ship was on the point of sinking. It seemed stupid to
let her cargo and provisions go with her, so they looted the ship,
then, in revulsion, tipped the body of Miller into the sea. They
then sailed away, expecting *Joyita* to sink and, because of their
guilt, did not own up when she was found.

It is a dramatic story, but, as Maugham would surely be the
first to admit, it is pure conjecture and, even so, does not entirely
fit. The Pacific islanders seem to have a particular gift for survival. In July 1962, 12 islanders out of a crew of 16 survived for
over three months in the half-submerged wreck of a Japanese
fishing boat stranded on a reef. Two more out of three survived
the attempt to fetch help in a homemade outrigger. But why
should the injured Miller be left behind anyway? If the Japanese

only wanted to help, why should Tanini fight them? Certainly it seems that a wave did crash onto the ship, that someone was injured and that the ship, when found, probably *was* looted by Japanese fishermen. All these assumptions accord with the few known facts. But the facts also add up to a much simpler theory— that Miller was injured when the wave fell on the ship and smashed down on the deckhouse and superstructure. The little bit of surgery was then performed, but because Miller was no longer able to tell everybody how safe he knew *Joyita* to be, the Europeans simply got everybody, including the unconscious Miller, into the rafts, overruling Simpson, McCarthy and Tanini, and taking the money and the ship's chronometer. When the Japanese arrived the vessel was already deserted, while the people in the rafts were already dead or dying from drowning, starvation or dehydration.

Or were they? In June 1958 six human skeletons were found in a cave on uninhabited Henderson Island, located about 100 miles from Pitcairn Island, where Fletcher Christian took the mutineers from *Bounty*. They appeared to be of recent date and the possibility was considered that they were the survivors of *Joyita*'s people. Officials at Pitcairn requested that they be placed in airtight coffins and transported to London or New Zealand for examination and identification, but the British authorities, seeing no value in such an operation, turned the suggestion down. A fragment of hair from one of the skeletons was examined by a pathologist at Suva, but it proved inconclusive, though the story is still believed that six of the people from *Joyita* were captured and taken aboard some vessel to Henderson Island, where they were marooned and died.

It seems unlikely, however, because Henderson is almost 2000 miles from the Fiji group, where *Joyita* was found. Dr. Edward Hunt, anthropologist at Harvard University, admitted it was a possibility, though he felt the skeletons were those of Polynesians, and it is a fact that Pacific caves were often used for the burial of native dead. Nothing was proved and the mystery remained as deep as ever.

As has often been stated of other sea mysteries, had there been just one survivor there would have been no mystery at all, and all the elaborate theories would have disappeared behind what was a surprisingly simple solution. The only thing that can be accepted with certainty is the anguish and terror of those aboard *Joyita* in the darkness of the night and the storm, without engines or a radio to call for help.

7. Teignmouth Electron (†1969)

A tragedy of loneliness

Disasters at sea have various causes, sometimes even the character of the man in command. Masters, like Captain Carey of *Vestris,* have proved inadequate, mates have been inefficient, crews or passengers have panicked. Without doubt, many people died in the *Titanic* disaster because officious stewards refused to permit steerage travelers to go to the first-class deck, and more still died because the fear that the suction created by her sinking would drag down the boats kept rescuers from picking up survivors in the water until exhaustion and the cold had accounted for many of them.

Without doubt, Miller's character had its bearing on the *Joyita* disaster, and once more, in the case of *Teignmouth Electron* in 1969, the drama was caused as much as anything else by the character of the only casualty.

The mystery began on the morning of July 10, 1969, when the Royal Mail vessel *Picardy* of the Furness Withy Line, bound from London for the Caribbean, sighted a small yacht ghosting along at about two knots with only a mizzen sail raised. The time was 7:50 A.M. and *Picardy* was then some 1800 miles out in the Atlantic in a position 33 degrees 11 minutes N, 40 degrees 26 minutes W—a place in the ocean where finding a small sailing yacht was unusual, to say the least. Chief Officer Joseph Clark roused Captain Richard Box, the master, from his bunk to take

a look at her. There was no one on the yacht's deck, and, pre-
suming the crew to be asleep, Box altered course to pass around
her stern and sounded the foghorn three times. There was no response. It was then possible to see the name of the yacht, a
trimaran—*Teignmouth Electron,* of Bridgwater. Thinking someone might be ill aboard her, Box had the ship's engines stopped
and a boat lowered. Clark, with a crew of three, investigated. He
climbed aboard and put his head into the cabin, but the boat appeared to be deserted.

Though the yacht was abandoned, she was in reasonably good
shape. The cabin was untidy, with two days' dirty dishes in the
sink, while three radio receivers, two of them disemboweled,
stood on tables and a shelf, with their parts strewn everywhere.
To one side a soldering iron was balanced on a can of milk,
which seemed to suggest that at least since being abandoned the
boat had not hit bad weather. An old dirty sleeping bag lay on
the forward bunk, the food and water supplies seemed adequate
and the equipment was in reasonable order, though the chro-

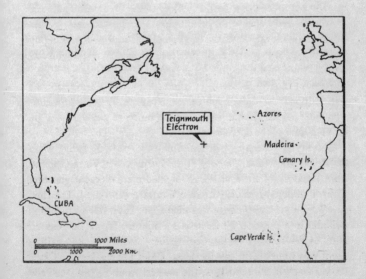

7. TEIGNMOUTH ELECTRON (1969)

nometer was gone from its case. The smell in the cabin seemed to indicate it had been empty for some days. On deck, a life raft was firmly lashed in place and the helm was swinging freely. Lowered sails were neatly folded, ready for immediate use. There was nothing aboard to indicate what had happened.

There were three logbooks in a neat pile on the chart table. They had been methodically kept, the last entry being for June 24, a fortnight earlier. The last entry in the radio log was for June 29.

Then someone remembered that *Teignmouth Electron* was one of the boats in the Golden Globe singlehanded nonstop race around the world organized by *The Sunday Times* of London. An old *Sunday Times* was unearthed and from it they learned that the crew of *Teignmouth Electron* was one Donald Crowhurst, of Bridgwater, Somerset, who had left Teignmouth, Devon, on October 31, 1968, the last man to start. They remembered he had been reported as putting up a splendid time through the Roaring Forties and in fact had been the only competitor left in the race.

A heavy derrick was rigged and the yacht was hoisted aboard. When the mast had been taken down and laid on the forward cargo hatch, Captain Box sent off a cable to his ship's owners in London, who passed the information to Lloyds and the Royal Navy, who asked the United States Air Force to start a search of the area. By then it was 10:30 A.M. local time and early afternoon in London. Box set lookouts and told them to look for a swimmer, but there was little hope because the dates in the logbook suggested that since Crowhurst had left the yacht over 10 days before, he could not have survived so long. Immediately the press, picking up the story from Lloyds, began to radio for details about what had happened, but Box would advance only the cautious reply that the whole thing was a mystery.

Next day *Picardy* and the Americans abandoned the search, and as *Picardy* continued toward Santo Domingo, in the Dominican Republic, Box began to read the logbooks in the hope of finding the answers to the puzzle. The messages in the radio log were carefully written down, and the navigation seemed precise

216

and seemed to indicate a successful voyage.

The tragedy seemed twice as terrible because, after a bad start, Crowhurst appeared to have been making tremendous times. He had even been catching up with the yacht ahead of him when that yacht had sunk, leaving him alone in the race and apparently certain to pick up the £5000 prize for the fastest circumnavigation. The prize for the first man home had already been won.

The first news of the boat's discovery reached Crowhurst's wife on July 10. At first she wasn't overly worried, thinking that her husband was in his dinghy. She asked if his wet suit was aboard in case he was swimming somewhere underwater, and someone else asked whether contaminated food had made him ill. Captain Box confounded both theories. The food was in good condition and the wet suit was still aboard.

As it became more obvious that a tragedy of some sort had occurred, two days later *The Sunday Times* started an appeal fund for the family with a £5000 donation, and Robin Knox-Johnston, who had been first home and as the only man to complete the course was now also eligible for the prize for the fastest time, generously asked that his £5000 prize should also go to the appeal.

Those people interested in the Bermuda Triangle mystery were quick to point out that *Teignmouth Electron* had been found in that area and came up with the information that four other vessels had been lost there in 11 days, while Lloyds were reported to have commented, "It's rare to get reports like this in such a close area in such a vast ocean. It's rather odd." In fact, one of the vessels reported lost was much nearer to Africa than where *Teignmouth Electron* was found, one was probably manned, one had probably been floating around abandoned for a long time and the fourth was never explained. And as it turned out *Teignmouth Electron* was a long way from the Bermuda Triangle anyway.

There were the usual references in newspapers to sea monsters and to *Mary Celeste,* because the circumstances were uncannily similar. There was a great surge of sympathy for Crowhurst and his family which, however, did not stop the usual legends from getting under way.

One was that Crowhurst was alive in South America and the *Daily Express* even sent a reporter to look for him in the Cape Verde Islands. The inevitable message in a bottle, this time found on a French beach, which said, "Help, stranded on island in the Aegean," was not even seriously considered. Later a photographer claimed to have seen Crowhurst in Barnstaple but admitted afterwards that perhaps he had been hoaxed. Other theories were that he had swum to the nearest land—a mere 700 miles!—or had set foot ashore secretly in the Azores and pushed off his boat, which had then managed to sail herself against wind and current to the point where she was found. One other theory, more likely than any and based on the fact that Crowhurst was known to be clumsy aboard boats, was that he had fallen overboard. The missing chronometer seemed to argue against this, however.

Crowhurst's voyage had been far from easy, because he had been beset all along with difficulties and had finally left England with only hours to spare before the deadline which would have disqualified him. He had been towed out of Teignmouth harbor, watched by friends and BBC cameras, still fuming because his halyards had fouled and his sails had been attached to the wrong halyards. After a fortnight at sea, he had seemed to be making very poor progress and had sent no radio messages. By this time he had sailed some 1300 miles but had progressed only a matter of 800 miles on his route. According to a radio message he sent, he was near Madeira, but then, on December 10, he left the prevailing southwesterlies and began to make better time. On that day he claimed to have broken a record by sailing 243 miles in one day. Always looking for a story, the press was delighted and began to see him as a possible winner, especially since the other competitors had begun to fall by the wayside. Chay Blyth and John Ridgway, who had rowed the Atlantic together, had already had to drop out and now others also began to fail. Alex Carozzo, known as the Francis Chichester of Italy, had retired because of a stomach ulcer. Bill Leslie King had just been towed into Cape Town with a smashed mast after turning turtle. Robin Knox-Johnston, in an ancient boat which was the smallest entry of the lot, was off New Zealand but his boat had been badly battered. He had already broken Francis Chichester's distance record for a nonstop voyage, but his self-steering gear had been smashed when his boat had capsized and his rudder was in disrepair. The Frenchman Bernard Moitessier, a famous figure among long-distance sailors and the author of two classic books about the sea, was closing the gap on him, while Nigel Tetley, also in a trimaran, was nearing the Cape, having already broken the record for the longest voyage in a multihull. Loick Fougeron, another Frenchman, in a well-tried steel-hulled boat, had battled his way through the Roaring Forties, but a capsize had left his boat a wreck and he was heading for dry land. So much had happened in a matter of days that all the betting on the race had changed as the favorites disappeared and the outsiders came up. As *The Sunday Times* said, in an article on this stage of the race, this was "the week it all happened." With Crowhurst suddenly making such splendid time it seemed that the vision of him as a possible winner was not so silly after all, despite the difficulties he had had to overcome before setting off.

Optimistic newspapermen, picking up his messages, added to the optimism felt by his supporters. They soon had him in the Indian Ocean, though the positions Crowhurst gave were vague; and by March 8, 1969, he even seemed to be narrowing the distance between himself and Tetley. With Knox-Johnston's battered little boat not expected to complete the voyage, Crowhurst began to look a likely contender for a prize because by this time Moitessier had decided he preferred sailing to going home and dropped out of the race, happily giving up both the prize and the Légion d'Honneur which was awaiting him if he won. This left only Knox-Johnston, Nigel Tetley and Crowhurst, and the race then seemed to have developed into a duel between the two trimarans because Knox-Johnston was believed to have been lost rounding the Horn. In fact, he was back in the Atlantic and was driving his battered little *Suhaili* north. On April 6 he was spotted at last by a tanker and celebrations were prepared for his return. But, though he would be first home, his time was very slow so that it seemed certain that one of the trimarans would win the £5000 prize for the fastest time. And by March 8 it began to look as though Crowhurst, despite his late start, was going to make it a very tight finish because he was narrowing the distance between himself and Tetley all the time. On March 20 Tetley battled his way around the Horn, but Crowhurst's messages seemed to suggest he was near Diego Ramirez, just to the south of the Horn.

Tetley's supporters were staggered by Crowhurst's miraculous reappearance and began to encourage their man to cram on speed. Then, on May 21, only 1200 miles from the end of his 30,000-mile voyage. Tetley's boat sank. Urged on by his supporters, he had tried too hard through the gales and his boat had begun to fall apart. On May 21 the port float bow broke away and smashed into the center hull and, as the boat filled with water, Tetley was obliged to take to his rubber raft.

It seemed then only a formality for Crowhurst to pick up the prize. All he had to do was reach home, taking no risks. At Teignmouth preparations were made for his return. He was to be escorted up-Channel, towed along the seafront and greeted on shore by civil dignitaries, watched all the time by BBC and ITV cameras. Even those who had been loudest at deriding his apparently amateurish efforts when he had set off were now grudgingly admiring, and those firms who had given him supplies were preparing advertisements to acquaint the public with the fact.

On June 18 the BBC sent congratulations and asked him to have his films and tapes ready, adding, "Any. suggestion on getting them at least four days before Teignmouth arrival? Can arrange boat or helicopter. How close Azores, Brittany or Scillies? Reply urgently."

On June 25 *Teignmouth Electron* was sighted by a Norwegian cargo ship, *Cuyahoga*. Crowhurst, bearded and apparently fit, waved cheerfully. The trimaran was on a northerly course, her position 30 degrees 42 minutes N, 39 degrees 55 minutes W, or about 750 miles southwest of the Azores.

His signals were still cheerful, even jovial. On June 26 he

received a message from his press agent to the effect that the BBC, the *Daily Express* and his wife were meeting him off the Scillies. It added that there were likely to be 100,000 people in Teignmouth to welcome him home and that the fund on his behalf had reached more than £1500, plus other benefits.

Crowhurst radioed back cheerfully, giving his position, but, curiously, insisted that his wife should not come out to meet him as had been suggested. On June 30 he made contact again. From then on nothing was heard from him, but because there had been long silences before no one worried. Perhaps he was too busy repairing things that had broken or was worn out by the long struggle. Only his wife was concerned about the change the around-the-world trip might have made in him.

Then on July 10 came the unexpected message from Captain Box of *Picardy*. *Teignmouth Electron* had been found deserted 250 miles from where it had been sighted by *Cuyahoga*. Donald Crowhurst had vanished. At the very moment when victory seemed certain, he had either fallen or been washed overboard.

His supporters were shattered, but perhaps there was still something to be salvaged from the disaster. Crowhurst's press agent, Rodney Hallworth, knowing the logbooks had been found, sold the copyright to *The Sunday Times* for £4000 and a group was organized to fly out and collect them. It included Hallworth and Nicolas Tomalin, of *The Sunday Times,* and as they flew over the spot where the boat had been found, Hallworth was sufficiently moved to ask them to maintain a moment's silence.

When they arrived aboard *Picardy,* they were still obsessed with the tragedy of what had happened to Crowhurst after such a magnificent fight, but then Box quietly called Hallworth to his cabin. Indicating the logbooks that Chief Officer Clark had found, Box said he had discovered from them that something was seriously wrong and that Crowhurst might even have committed suicide. He urged the startled Hallworth to tear out the relevant pages so that his family should never know. Hallworth did so, intending to show them only to the editor of *The Sunday Times,* and it was only the next day, after carefully examining the logbooks himself, that he made another discovery which he felt he could no longer keep to himself, and he read the last words in the log to his companions.

What Hallworth had discovered was a sensation, but not the sensation he had expected. *The Sunday Times* found itself with the scoop of the year, but because there had been much talk about newspapers setting up dangerous stunts without proper supervision, they were far from happy about it. When the party arrived back in England, an editorial conference tried to decide what to do. Crowhurst had left an extraordinary story in his logbooks which made reasonably clear what had happened to him, and it didn't seem to reflect a great deal of credit on the sponsors of the race. Mrs. Crowhurst also had to be considered, but the story could clearly not be suppressed and the courageous decision was taken to publish a full account.

Mrs. Crowhurst was told and the BBC, who were preparing a version of Crowhurst's heroic voyage, were asked to delay their program. Only the Director General was informed why.

The story when it appeared *was* a sensation. What Captain Box had discovered by reading through the logbooks was that Crowhurst had never rounded Cape Horn at all—or for that matter the Cape of Good Hope! He had spent the entire eight and a half months he had been away from England wandering about the Atlantic. Far from putting up the gallant fight he seemed to have put up, he had faked his radio messages and positions, and Fleet Street had fallen for them hook, line and sinker.

The celebratory Golden Globe dinner was put off and, to his enormous credit, Knox-Johnston still insisted on giving his £5000 prize to the Crowhurst appeal fund while Tetley received a £1000 consolation prize. Knox-Johnston's comment, "None of us should judge him too harshly," was not only generous but also very wise. The tragedy of *Teignmouth Electron* was entirely one of personality.

The strain of eight and a half months' constant and lonely vigil had been tremendous. Both Francis Chichester and Knox-Johnston had been aware of it, but they were both mentally tough and had already had experience of lonely command. They were well fitted for what they had done. Crowhurst's character seemed to suggest that he wasn't, and his personality, perhaps never very stable, had cracked under the strain of the tremendous deception he had been practicing.

Donald Crowhurst was born in India in 1932, his mother a schoolteacher, his father, John Crowhurst, a superintendent on the railways. Despite a tendency to drink too heavily at times, John Crowhurst was a highly intelligent man and his son Donald was self-assertive and brave. In 1947, at the time of the partition of India, Crowhurst brought his family to England and put his savings into a small sports goods factory in Pakistan which was to be run by a Pakistani partner.

They settled at Tilehurst, near Reading, but found it very different from India. There were no servants and Crowhurst even had to take a job as a porter in a jam factory, while his factory in India was burned down in the partition riots. His son had to leave school as soon as he had passed his school certificate and very soon afterwards, in 1948, the father died.

Donald Crowhurst was accepted for the Royal Air Force and studied electronic engineering at Farnborough Technical College. During the next six years he passed his technical examinations, learned to fly and was commissioned. Suddenly with money in his pocket, he bought a secondhand Lagonda. He began to show a madcap wish to defy authority, and was always inclined to show off and take risks. It did not go down well and in the end he was asked to leave the RAF—because, it was said, he rode a powerful motorbike through barrack rooms full of sleeping men at Farnborough. Another story was that he missed a vital parade to try his hand with the Lagonda at Brands Hatch. Whatever it was, it was not serious enough to prevent his being immediately commissioned in the army and taking another course in electronic control equipment. His temperament hadn't changed, however, and he continued to self-dramatize. Once again he took risks and was caught driving his car without insurance. Eventually in 1956 the army also asked him to resign.

At loose ends, he decided to try for Cambridge. His place was secure but he failed the Latin qualifying exam. Doing research work in the Reading University laboratories, at 24—considered rather dashing and a bit of an intellectual with a definite talent for achieving things—he met his wife, Clare, who was Irish, at a party in 1957. They married in October of that year and he joined Mullards, an electrical components firm, but it turned out not to be the research job he had expected but merely that of a specially qualified commercial traveler. When he had a bad crash in a company car and was reprimanded, he threw up the job. He was now 26 and, rather like Miller of *Joyita,* everything he had tried had collapsed. He worked in Maidenhead and then for a Bridgwater electrical engineering firm, all the time determined to set up on his own. He was certainly skilled in his field and he enjoyed presenting an image of himself as an inventor of electronic gadgets.

About this period he bought a small 20-foot sloop which he kept near Bridgwater, and his interest in sailing helped him to produce a well-designed if not entirely original navigational gadget called the Navicator. By then he had a growing family of three boys and a girl. He was popular, vivid and clever, but in a small place like Bridgwater there was no one equally clever who might have disciplined him, and though in his own field his ideas were impressive, he always lacked intellectual control. His writings were marred by misspelling and unexpected inconsistencies, while as a mathematician he could ruin an inventive idea by faults in simple addition. His reaction to failure was always depression, though it rarely lasted long and would always eventually erupt into high spirits. He was eager for success, to prove that he was an achiever, but he had not yet learned to wait.

In 1962 his mother tried to kill herself and was put into a hospital. Her house was sold and, with her consent, her son used some of the money from the sale to launch his firm, Electron Utilisation, to market his Navicator. He tackled the job with his usual energy and it seemed set for success. Soon afterwards he became a Liberal member of the Bridgwater Borough Council, claiming that his business management skills could help solve the town's industrial problems.

He bought a new Jaguar but drove it too fast, and when he crashed, the head injury he received seemed to affect him, leaving him moody. He became interested in the supernatural, something his wife didn't like, but his firm still looked a success and he was employing six full-time workers. Pye Radio even began takeover negotiations and paid him £8500, but the deal fell through. Things began to go wrong so that, short of capital, he had to leave his rented factory and manage with one part-time assembler.

He then met a man named Stanley Best, a Taunton businessman who sold trailers. Best put money into the firm, though he soon discovered that although Crowhurst was a brilliant innovator he was not very good with the hard matter of finance. "In a workshop," he said, ". . . he was superb. But as a businessman he was hopeless. He seemed to have this capacity to convince himself that everything was going to be wonderful, and hopeless situations were only temporary setbacks." His enthusiasm was infectious but in reality it was the result of overimagination and a tendency to see things as he wanted them to be.

It was at this point in Crowhurst's life, in 1966–67, that Sir Francis Chichester made his solo around-the-world voyage. He was knighted as a result and around-the-world sailing caught the public imagination. It was not new, because Joshua Slocum had done it first in 1895–98 and others had followed, but *The Sunday Times,* which had supported Chichester, encouraged the fever. It became Crowhurst's one ambition to repeat the performance. He probably even began to see it as a means of saving his failing firm, because Chichester had made a lot of money in royalties and in lecture and TV fees. By the end of the year at least four yachtsmen had decided to cap Chichester's achievement. Bill Leslie King was an ex–submarine commander with a boat smooth and rounded enough to look like a submarine, but with two junk-rigged masts. Robin Knox-Johnston, a 28-year-old, extraordinarily stable-minded ex–merchant navy officer, was about to attempt the voyage in *Suhaili,* a tiny 32-foot teak-hulled Bermudan ketch he owned. To the experts, *Suhaili* seemed far too small and vulnerable, but she had sailed from India and was easy to handle. There was also Bernard Moitessier, a legendary Frenchman who had sailed thousands of miles in the Pacific, and finally John Ridgway, formerly of the SAS but of no particular distinction as a yachtsman, who applied for leave to make the attempt. Crowhurst became obsessed with the idea that he must try, too, and did his best to obtain Chichester's *Gypsy Moth IV*. He failed and, as it happened, Chichester decided he didn't like the look of Crowhurst, and his suspicions never altered.

Crowhurst then approached Lord Thomson, owner of *The Times,* suggesting that one of his papers should sponsor an around-the-world nonstop voyage or even perhaps a race. By a coincidence the idea had already been suggested from an entirely different source. When it was announced by *The Sunday Times* two weeks later Crowhurst claimed it was at his suggestion.

What had happened, in fact, was that *The Sunday Times,* having gotten wind of the enthusiasts preparing to tackle the lone circumnavigation of the globe, had decided on a race in which the competitors could start when and where they liked. They did not even have to enter, simply have their departure and arrival recorded by a national newspaper or magazine. There would be a prize for the fastest trip and one for the first home. Since the newspaper would be unable to vet the competitors, who would all be starting from different points, an expert panel of judges was set up, including Sir Francis Chichester.

There was now another competitor, an Australian dentist named Howell, who had a trimaran and was strongly tipped to win. Several other would-be competitors, including a youth with a homemade boat, had to be talked out of it, though Chay Blyth, who joined after only a few days' experience, was considered to be able to look after himself since he had helped John Ridgway to row across the Atlantic. When the race was announced in March 1968 Crowhurst also announced himself as a competitor, and, though he did not yet have a boat, since he had convinced a great many people that he ought to have Chichester's *Gypsy Moth IV*, he was not turned down.

By May the first two, Blyth and Ridgway, were almost ready to leave, as was Knox-Johnston, while a second expert Frenchman, Loick Fougeron, had joined the race with a steel-hulled boat. Crowhurst had decided to sail in a trimaran, despite the fact that they were not popular with singlehanded sailors who felt that the helm needed to be continuously manned and that, though fast before the wind, they were poor performers to windward. And though they didn't capsize easily, when they did they were virtually impossible to right.

It was at this point that Crowhurst had a little luck. Stanley Best, worrying about the money he had put into Crowhurst's failing firm, tried to pull out, but instead Crowhurst talked him into financing the building of a trimaran. Best's wife thought he must have gone mad. "It was, I suppose," he admitted, "the glamour of the idea, the publicity and the excitement—and the persuasiveness of Donald. He . . . was the most impressive and convincing of men."

Crowhurst gave him the impression that others were helping with the expense and agreed that if anything went wrong the boat would be bought by his firm, Electron Utilisation. He was very plausible. The boat would be equipped with electronic devices of Crowhurst's invention. "If the practical utility of the equipment I propose," he wrote, "can be demonstrated in such a spectacular way as in winning *The Sunday Times* Golden Globe race . . . and it is properly protected by patents, the rapid and profitable development of this company cannot be in any doubt."

The hulls for his boat were built at Brightlingsea, Essex, and assembled by L. J. Eastwood Ltd. of Brundall, Norfolk, who agreed to do the job for a minimum profit. As the designing was hurried ahead, it was found that another trimaran had been entered by Commander Nigel Tetley, whose boat was a Victress-class vessel large enough for him and his wife to have lived on it for several years. Crowhurst's designs were based on this class of boat and he expected to sail on October 1. He had no doubts about winning and he even worked out the order of finishing, placing the other two multihulls in second and third places.

By this time he was making decisions about the boat on which his life might depend. He had, he said, found a means of overcoming the trimaran's chief drawback—its tendency to remain upside down if capsized—and had built an electronic system which would not only give early warnings of unusual stresses in the rigging and slacken the sails, but, in the event of a capsize, would right the boat by means of a computer which would fire off a carbon dioxide cylinder attached to the mast. In fact he had not yet even designed it.

Because of all the electronic gadgetry he claimed to have prepared, Crowhurst insisted on a large generator to work them, but he took what proved to be a faulty decision to place it below the cockpit. There were other risky decisions, chiefly because both Crowhurst and the boatyard were working against time and because Crowhurst was also trying to take a radio-telegraphy course. He had acquired a press agent, Rodney Hallworth, formerly of the *Daily Mail* and the *Daily Express* and by this time proprietor of the Devon News Agency. But chiefly because he had left it until too late, he could find no firm who would be willing to sponsor the trip, so that Stanley Best found himself footing the bills for most of the equipment, though the BBC, who bought the TV and tape-recording rights, supplied a camera and a tape recorder. The boat became known as *Teignmouth Electron*.

Unfortunately work was falling behind schedule and she was not launched until September 23, only just over a month before the deadline. By this time everybody was at loggerheads because of the rush and the conflicting instructions. Meanwhile Howell, disappointed in his boat's performance in the *Observer* Transatlantic Race, had dropped out, though several of the competitors were already on their way and there was a constant need to hurry.

It was hoped that the maiden voyage of *Teignmouth Electron* from Brundall to Teignmouth would take three days. In fact it took two weeks and delays caused the crew to be twice changed, so that Crowhurst ended up at one point sailing it on his own. Unhappily the trimaran proved as difficult to windward as people claimed such boats were, and during this period, while talking about its failings, Crowhurst was asked what he would do if he failed to force the boat down the Atlantic to the Southern Ocean. "Well," he said, "one could always shuttle around in the South Atlantic for a few months. There are places out of the shipping lanes where no one would ever spot a boat like this." He then showed how it could be done, tracing a course around and around the Falkland Islands and Tristan da Cunha. It would be simple, he said. The idea was accepted merely as a joke.

By this time Blyth and Ridgway had dropped out of the race, but Knox-Johnston was in the Indian Ocean and approaching Australia, while Moitessier, King and Tetley were chasing him down the Atlantic. At Cowes Crowhurst met a late entrant for the race, Alex Carozzo. He had a huge 66-foot ketch and Crowhurst was sufficiently impressed to consider him his most dangerous rival. At Cowes he also took on a new crew member, Peter Eden, who noticed his clumsiness on board and his tendency to fall into the water. The trimaran, though it could not get close to the wind, appeared to sail very fast, though at speed the vibration was so violent the screws holding the self-steering gear kept working loose so that the crew had to keep leaning out over the counter to tighten them. Eden advised that the fixings should be welded if the gear were to survive. He also realized that although Crowhurst was a good sailor, he was inclined to be slapdash at navigation.

They arrived at Teignmouth only 16 days before the deadline and with a great deal still to do. With time running out, at times Crowhurst seemed in a panic, at others stupified by all that still had to be done. Yet at others he seemed full of confidence. Even at this stage, however, the BBC man in charge of shooting the preparations began to suspect they were shooting not for a potential triumph but for a potential tragedy, and on the eve of departure, the rush was so great they even stopped filming and helped by buying and putting aboard necessities. Aware of the problems, a friend told Crowhurst more than once that he shouldn't go, but he always replied that it was too late to turn back. According to his wife, he was already aware that there could be no success and, afraid to admit he had failed, was praying that she would beg him to back out of the project. Unfortunately, his apparent confidence utterly misled her and, unaware of his growing misery, she could only try to encourage him.

Crowhurst finally managed to leave with nine hours to spare

but with supplies and spares strewn everywhere and things hopelessly wrong or missing. Instead of doing what Carozzo did— complying with race rules by setting sail and then anchoring until he had stowed his gear, checked his equipment and made sure all was in order—he headed straight toward Cape Finisterre. From this point on, when he was entirely alone, what he thought and did can only be discovered from the logs that Captain Box read aboard *Picardy* eight and a half months later.

Crowhurst was already in despair. Confined to a tiny cabin, alone, the electronic wizardry he had boasted about nothing more than wire and unassembled parts, he had not even had time to plan a course. His steering gear was still shedding screws which he had to keep replacing with screws taken from other parts of the boat. "This bloody boat is just falling to pieces," he wrote in his log a fortnight after leaving. It was a terrible prospect. He was afraid the wild broaches caused by the self-steering gear might well capsize the boat, and he had no radio communications and none of the masthead buoyancy he had boasted about. In addition, he had discovered that the port bow float was leaking, yet he had no means of pumping it out because a pipe he needed had not been put aboard, and with the cockpit hatch letting in water, the engine compartment had flooded so that the generator, placed low down in the hurry of final decisions, had been stopped. Now he did not even have electricity, and added to all this were leaky hatches, incorrectly cut sails, unsatisfactory shrouds and stays, and bad stowage arrangements. For a long time he seriously considered dropping out or merely going as far as Cape Town on a face-saving trip.

By the time he got the generator going and sent off a message that he was heading toward Madeira—exaggerated by the press to "near Madeira"—his position seemed hopeless. In a radio-telephone message to his wife, however, he concealed his doubts, but, paralyzed now by indecision, he had come almost to a dead stop as he struggled to decide what to do. Racked by a growing awareness that he had to decide whether to continue, he was in a dreadful dilemma. After all he had said and done, the idea of dropping out was a hateful one, but, with all that was wrong with his boat, he was also well aware that he would be taking a dreadful risk if he tried to round the Cape or the Horn.

But then, as he struggled with himself, he found he had left the prevailing southwesterly winds behind and he began to do better. It was around this point that his deception started and he began to keep two logbooks, one true, one false. His increased speeds seemed to go to his head and his claim, made by radio on December 10, to have set a record of 243 miles in one day delighted his supporters. "LONE SAILOR CLAIMS A NEW RECORD," was the headline in the *Daily Mirror,* which stated he was now south of the equator and heading toward Cape Town. *The Sunday Times* decided, "The achievement is even more remarkable in the light of the very poor speeds in the first three weeks of his voyage." Sir Francis Chichester, however, was not deceived and considered Crowhurst "a bit of a joker" who would need watching.

By this time Crowhurst had started to write down fraudulent sun sights to make the log sound convincing, but at the same time he was also beginning to worry about his false claims and, wondering whether at the end of his trip he could face such experienced men as Chichester, he started to plan a fake voyage. The radio positions he gave were always ambiguous.

About this time too he discovered his starboard float had split and he was also beginning to realize that if he failed, under his agreement with Best he was facing bankruptcy and now even exposure as a cheat. Nevertheless, his messages were always being read too optimistically at home, so that he was always considered to be farther ahead than he was, and he was more and more caught up in the web of his own making.

Though the tape recordings he made for the BBC were full of confidence, by Christmas his log showed the strain he was undergoing, lonely, full of introspection and self-doubt, pottering along with no clear destination, worried about his boat and growing more and more aware that some time soon he would have to get his leaking float repaired. By this time he was taking copious notes of broadcasts to shipping, cable traffic, news bulletins and weather reports, all so that he could record in the false log weather conditions in the areas he was supposed to be traversing.

His own messages home grew more and more vague, and newsmen always helped the deception by reading far too much into them. Only Chichester remained unconvinced and was even suggesting that Crowhurst's claims would have to be checked.

Confined to a tiny cabin and deck space, in his silence and loneliness, Crowhurst grew more and more miserable and took to writing poems and long prose pieces as he worried about the urgent need to repair the leaking port float. The split which was causing the leak was increasing and, in trying to keep it out of the water, he had been unable for some time to steer any kind of useful course, while, thanks to the last-minute mixups before he had left, he had no means of repairing it. By putting ashore, however, he could be disqualified and certainly would risk its being discovered that he was nowhere near where he claimed to be.

For a long time he struggled with his thoughts before finally, in desperation, he headed for the lonely Rio Salado in Argentina, where he dropped anchor on a falling tide on March 6. Immediately he found himself aground and, when he was towed off by a local boatman, he broke the rules of the race for the first time. Up to then his voyage had been wandering and unorthodox but he had broken none of the rules. Now he had to think of that, too.

To the Argentines he appeared to be thin but healthy, but he talked disconnectedly and strangely. The only man who might have exposed him was a midshipman at the La Plata Prefecture, to whom the local coast guard petty officer telephoned for a decision about giving help. This midshipman did not even consider it important enough to inform his seniors, and Crowhurst left Rio Salado on March 8, after repairing his boat, with no one beyond the immediate vicinity aware that he had ever set foot ashore.

Perhaps it was this contact with human beings after being alone for so long which was the last straw. Chichester had noticed on his around-the-world trip that it became harder to be alone after having spoken to people in Australia. Crowhurst was suffering from the strain of the lies he had told, the false log he was keeping and the new fear that his landing might be discovered, so he sailed farther south in the hope of shooting pictures of the Horn which would act as proof that he had been around it.

By this time, with Carozzo, King, Moitessier and Fougeron out of the race, only Knox-Johnston and Tetley were left and Crowhurst was placed by newsmen at home as lying close behind Tetley. He was lonely, depressed and increasingly obsessed with thoughts about God and the universe, so that his writings became increasingly rambling.

With Knox-Johnston safely home, Crowhurst was now trying hard *not* to beat Tetley. From what can be read in his logbooks it is clear that he was hoping to be accepted as a good loser, because that way he would be hailed as a hero while there would be no need to check his logbooks. Though he was going through the motions of racing, he was actually only dawdling in the hope that Tetley would reach home first.

When Tetley's boat sank he found himself in a dreadful dilemma. It was the supreme irony. Because Tetley's supporters believed Crowhurst's messages that he was close behind and urged Tetley to take the risks that finally sank his boat, Crowhurst made it inevitable that if he continued, the prize would be his. Now, he realized, there would be a hero's welcome, and with it there would also be a close inspection of his logbooks, something which terrified him because, despite his false positions, the experts would surely expose him as a cheat.

When his faked route had joined his actual position he had said he was on the way home and Hallworth had urged him to hurry. "Photo finish will make great news," he had cabled, and Crowhurst was worried sick. As the only competitor still sailing, it seemed he couldn't fail to win the prize for the fastest circumnavigation. But he was now in the full glare of publicity. His previous failures had always been concealed by changing jobs, but this time there was no chance of hiding it. He knew he would be exposed. Certainly Chichester was growing more and more suspicious. He was even demanding a list of his authenticated messages and positive position statements, and was asking why an electronics engineer should have so many radio silences at important times.

By June 24 Crowhurst's mind seems to have given way and,

while at Teignmouth the preparations were going ahead for his return, he was writing vague wandering soliloquys, full of under-linings, exclamation marks and capital letters. He had been spotted the day before by the Norwegian freighter *Cuyahoga*, apparently fit and well, sailing northeast, but in fact he had lost all touch with reality. The only thing he could think about was how to escape the consequences of his fake around-the-world voyage.

In his log he rambled on about God, Catholicism, the cosmic mind and the Devil, and had lost all track of time.

"I see what I am," he wrote, "and I see the nature of my offence. . . . It is finished—it is finished IT IS THE MERCY."

About this time he removed the line he normally streamed astern as a precaution against falling overboard and spent his time writing, desperately in need of help, desperately seeking a decision and some sort of forgiveness. The last three pages of his log indicated the terrible state of his mind. Among the scrawled philosophical meditations were odd cryptic phrases:

Nature does not allow God to Sin any Sins except one— That is the Sin of Concealment. . . . It is the end of my game. . . . The truth has been revealed and it will be done as my family require me to do it. . . . It is time for your move to begin. . . . I have no need to prolong the game. . . . It has been a good game that must be ended at the . . . I will play this game when I choose I will resign the game. . . . There is no need for harmful . . .

He stopped writing at 11:18 A.M. some time after June 29 and, piling the logbooks he had written on top of each other, he placed them near the navigational plotting sheets that showed his deception for the record run he had claimed. The dirty dishes were stacked in the sink and, judging by the tufts of it found on the deck of the cabin, he cut his hair. It is assumed that he then took up the chronometer and stepped off the stern of the boat into the sea.

Months after Crowhurst's disappearance, people were still insisting he was alive, having been taken off his boat in mid-Atlantic by a Spanish or Norwegian cargo ship whose crew he had sworn to silence. He was considered to be quite capable of doing it because he was remembered as a most persuasive man and, indeed, he had persuaded the hardheaded Stanley Best, instead of withdrawing his support for Crowhurst's firm, to finance another, even wilder venture, the building of a boat.

But whatever else he was, Crowhurst was neither a fool nor a villain. He had even shown great courage in sailing his dubious leaky boat into the Southern Ocean within 500 miles of the dreaded Horn. His achievement had not been small because he proved himself a first-rate helmsman. Captain Craig Rich, lecturer at the School of Navigation, who was advising the race organizers on this subject and who examined Crowhurst's logbooks, considered him a sufficiently skilled navigator to have made the around-the-world voyage. "The fact that he sailed over 16,000 miles singlehanded," he said, "must be considered a remarkable achievement in itself." Three years before, it would have been considered magnificent. Even before he put ashore at Rio Salado, after 8155 miles, he had sailed a multihulled boat singlehanded and nonstop farther than anyone save Nigel Tetley.

The ironic thing is that if Crowhurst had continued on a proper course around the world he might just have made it and come home, if not a winner, then the hero he wished to be. And as the yard that fitted out his boat said, there were many advanced features in his original conception that would then have made him money. Because of the rush, however, there was never time to develop and test them.

The tragedy of *Teignmouth Electron* was one that did not spring from any lack of skill or physical courage on the part of Crowhurst. It sprang from the fact that he didn't have the moral courage to admit he had made a mistake, and that he lacked the mental stability necessary for a lone long-distance sailor. He had the same attitude toward the truth, according to one newspaper article, as John Stonehouse and Richard Nixon, and convinced others because he convinced himself, something which led people to back him and support him not only in the electronics venture but finally in his disastrous voyage. He was a

paranoid, though stable, and throughout his life had shown a de-
lusional conviction about his own skill and ability.

Yet perhaps it was less the loneliness that finished Crowhurst
than the absence of a goal. Had there been something to strive
for, even merely getting around the world, let alone getting
around first, he might have survived. As it was, the rushed preparations had given him an unfinished, unprepared boat, and the
purposelessness of hanging around in the South Atlantic left him
too much a prey to his own thoughts. Despite the theories that
were advanced, there seems little doubt that Crowhurst's mind
went. He wrote 25,000 words, all rambling and disconnected,
something unbelievably difficult for a sane person to do. "Try
it," one psychiatrist suggested.

And, despite the extensive research of *The Sunday Times* men,
Nicolas Tomalin and Ron Hall, who covered the tragedy in a
restrained and sympathetic book, what happened at the end is
still a mystery. Though it seems he must have committed suicide,
there was still the evidence of his clumsiness aboard a boat when
preoccupied. His friends had often seen him fall into the water,
and there remains the possibility that this is what he did.

Sources

Armstrong, W. *Last Voyage*. London: Frederick Muller Ltd., 1956. New York: John Day Co., 1958.

Barnaby, K. C. *Some Ship Disasters and Their Causes*. London: The Hutchinson Publishing Group, Ltd., 1968. Cranbury, New Jersey: A. S. Barnes & Co., 1970.

Breed, Bryan. *Famous Mysteries of the Sea*. London: Arthur Barker Ltd., 1965.

Crocker, Sir William. *Far From Humdrum: A Lawyer's Life*. London: The Hutchinson Publishing Group, Ltd., 1967. Cleveland, Ohio: World Publishing Co., 1970.

Cyriax, R. J. *Sir John Franklin's Last Arctic Expedition*. London: Methuen & Co. Ltd., 1939.

Fay, Charles Edey. *Mary Celeste: The Odyssey of an Abandoned Ship*. Peabody Museum and Atlantic Mutual Assurance Co., 1942.

Freidel, Frank. *A Splendid Little War*. Galley Press, 1958. Boston: Little, Brown & Co., 1958.

Hastings, MacDonald. *Mary Celeste: A Centenary Record*. London: Michael Joseph Ltd., 1972.

Hawkey, Arthur. *H.M.S. Captain*. London: G. Bell & Sons Ltd., 1963.

Hoehling, A. A. *They Sailed Into Oblivion*. London: Thomas Yoseloff Ltd., 1959. New York: Thomas Yoseloff Inc., 1959.

———. *The Great War at Sea*. London: Arthur Barker Ltd., 1965. New York: Thomas Y. Crowell Co., 1965.

Hough, Richard. *Admirals in Collision*. London: Hamish Hamilton Ltd., 1959. New York: Viking Press, 1959.

Keating, Lawrence J. *The Great Mary Celeste Hoax*. Heath Cranton, 1921.

Kusche, Lawrence D. *The Bermuda Triangle Mystery—Solved*. London: New English Library Ltd., 1975. New York: Harper & Row Publishers, 1975.

Lockhart, J. G. *Strange Adventures of the Sea*. London: George Allen & Unwin (Publishers), 1929.

———. *The Mary Celeste and Other Strange Tales of the Sea*. St. Albans, Hertfordshire, England: Hart-Davis Educational Limited, 1952. Fair Lawn, New Jersey: Essential Books/Oxford University Press, 1952.

Lundberg, Ferdinand. *Imperial Hearst*. New York: Arno Press, 1927, 1970.

McKee, Alexander. *Death Raft*. London: Souvenir Press Limited, 1975. New York: Charles Scribner's Sons, 1976.

Markham, Captain A. H. *Life of Sir John Franklin*. London: George Philip & Son Ltd., 1889.

Markham, Sir Clements. *The Lands of Silence*. Cambridge, England: Cambridge University Press, 1921.

SOURCES

Maugham, Robin. *The Joyita Mystery*. London: Walter Parrish International Ltd., 1962.

Mielke, Otto. *Disaster at Sea*. London: Souvenir Press Ltd., 1958. New York: Fleet Publishing Corporation, 1958.

Mirsky, Jeannette. *To the Arctic*. Allan & Wingate, 1949. Chicago: University of Chicago Press, 1970.

Ortzen, Len. *Strange Mysteries of the Sea*. London: Arthur Barker Ltd., 1976. New York: St. Martin's Press, 1977.

Osborn, Captain Sherard. *The Career, Last Voyage and Fate of Franklin*. London, 1860.

Owen, Roderic. *The Fate of Franklin*. London: The Hutchinson Publishing Group Ltd., 1978.

Post, Charles Johnson. *The Little War of Private Post*. Boston: Little, Brown & Co., 1960.

Rickover, Admiral H. G. *How the Battleship "Maine" was Destroyed*. Washington, D.C.: Department of the Navy, 1976.

Ruhen, Olaf. *Minerva Reef*. Lewes, Sussex, England: Angus & Robertson (U.K.) Ltd., 1963. Boston: Little, Brown & Co., 1964.

Sigsbee, Captain C. D. *The Maine: A Personal Narrative*. Century, 1899.

Smith, G. W. and Judah, C., eds. *Life in the North During the Civil War*. Albuquerque, New Mexico: University of New Mexico Press, 1966.

Swanberg, W. A. *Citizen Hearst*. Harlow, Essex, England: Longman Group Ltd., 1962. New York: Charles Scribner's Sons, 1961.

Tolley, Rear Admiral Kemp. *The Cruise of the Lanikai*. New York: Naval Institute Press (dist. by Arco Pub. Co.), 1973.

Tomalin, Nicolas and Hall, Ron. *The Strange Voyage of Donald Crowhurst*. Sevenoaks, Kent, England: Hodder & Stoughton Ltd., 1970. New York: Stein & Day, 1970.

Villiers, Alan. *Posted Missing*. Sevenoaks, Kent, England: Hodder & Stoughton Ltd., 1974. New York: Charles Scribner's Sons, 1974.

Wright, Rear Admiral Noël. *The Quest for Franklin*. London: William Heinemann Ltd., 1959.

Newspapers, magazines and other sources: *Cape Argus; Cape Times; Century Magazine; Cornhill Magazine; Daily Express; Daily Mirror; Daily News*, Durban; *Fortnightly Review; Hampshire Gazette; Illustrated London News; Mariners' Mirror; Natal Mercury; National Star; Nautical Magazine; Portsmouth Evening News; Scientific American; Strand Magazine; The Sunday Times; Sunday Tribune*, Durban; *The Times; United States Naval Institute Proceedings*, 1969, 1970, 1973 and 1974; *Virginian-Pilot and Norfolk Landmark;* and the National Maritime Museum, Greenwich, England.

INDEX